THE MIRACLE OF FLIGHT

THE MIRACLE OF Flight

STEPHEN DALTON

FIREFLY BOOKS

A FIREFLY BOOK

Published by Firefly Books Ltd. 1999

First Printing 1999

Library of Congress Cataloging-in-Publication Data

Dalton, Stephen.
  Miracle of flight / Stephen Dalton – 2nd ed.
[176] p. ;   col. ill. ;   cm.
Includes index.
Summary : How animals and humans have evolved wings—natural and mechanical—revealing the magic of flight; new illustrations throughout.
ISBN 1-55209-378-6
1. Flight. 2. Aerodynamics. I. Title.
591.479–dc21   1999   CIP

Published in the United States in 1999 by
Firefly Books (U.S.) Inc.
P.O. Box 1338, Ellicott Station
Buffalo, New York 14205

Produced by
Bookmakers Press Inc.
12 Pine Street
Kingston, Ontario K7K 1W1
(613) 549-4347
tcread@sympatico.ca

Design by
Janice McLean

Printed and bound in Canada by
Friesens
Altona, Manitoba

Printed on acid-free paper

Canadian Cataloguing in Publication Data

Dalton, Stephen
  Miracle of flight

Rev. ed.
Includes index.
ISBN 1-55209-378-6

1. Aerodynamics. 2. Animal flight. I. Title.

TL570.D34   1999   629.13   C99-930379-1

Published in Canada in 1999 by
Firefly Books Ltd.
3680 Victoria Park Avenue
Willowdale, Ontario M2H 3K1

Canadä

The Publisher acknowledges the financial support of the Government of Canada through the Book Publishing Indus-try Development Program for its publishing activities.

ACKNOWLEDGMENTS

Dedicated to all my family

The author is most grateful to the following colleagues who have provided expert advice or have helped in other ways in the preparation of this book:

Professor Charles Ellington, a zoologist and microaerodynamicist, for reading the chapter on insect flight and allowing me to use the results of his latest research on how insects manage to get into the air at all.

George Hayhurst, for creating the meticulously drawn diagrams.

Captain Terry Henderson, for checking the final chapter, "The 20th Century and Beyond," and for his enthusiasm about this book. Terry's knowledge of aircraft is encyclopedic. Apart from owning his own Chipmunk airplane and Optimist sailplane, he has piloted the Concorde for some 25 million miles for British Airways.

Dr. David Newman, for allowing me to reproduce his scanning electron micrograph, which shows the beautiful three-dimensional structure of a damselfly wing.

Dr. Robin Wootton, for his guidance on the evolution of insect flight and the controversial "click mechanism." Robin has also been kind enough to draw me a protopterygote, a possible ancestor of the flying insect.

My long-suffering wife Liz and our lively children, who have supported the project from the start and have stoically put up with my occasional irascible outbursts yet have hung on my every written word!

And, finally, Bookmakers Press— editor Tracy Read, copy editor Susan Dickinson and art director Janice McLean, who have made such a superb job of producing this book. Although we have never met, we were in close contact through what seemed thousands of complicated E-mails and parcels of proofs, which winged their way across the Atlantic almost every day for six months.

Stephen Dalton

ARCHIVAL DISPLAY PRINTS

All the photographs in this book are available as high-quality archival display prints. For details, please visit www.nhpa.co.uk/dalton.htm

CONTENTS

INTRODUCTION

A Holistic Overview
8

CHAPTER 1

The Fundamentals
14

CHAPTER 2

The First to Fly
36

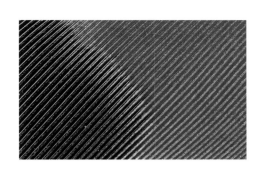

CHAPTER 3

The Feathered Wing
76

CHAPTER 4

The Evolution of Manned Flight
116

CHAPTER 5

The 20th Century and Beyond
144

FURTHER READING 172

INDEX 174

CREDITS 183

A Holistic Overview

Since the dawn of history, man could not fail to have been impressed by the mysterious phenomenon of flight. As birds and insects were constantly around for him to observe and admire, he must have longed for wings of his own so that he, too, might take off and experience the rush of cool air as he swooped and turned in the skies. Unobstructed and free, he would escape the constraints of his earthbound existence. The notion of flight would have seemed even more of a miracle then than it still appears to many of us today, so it is no wonder that wings, with their power, mobility and speed, symbolized the unearthly, the superhuman and the divine.

Wings are among the most extraordinary of the many complex structures that have evolved in the animal kingdom. Insects and birds between them make up well over three-quarters of all land creatures, and it is their ability to fly that has made them so unusually successful, allowing them to develop into a huge variety of forms and species living in diverse habitats all over the globe.

For man, too, the mastery of flight probably represents his supreme technical achievement, which, in less than a century, has enormously extended his domination of the planet.

Only in the past hundred years or so have we begun to understand the physical laws that govern flight, yet we are still moved or sometimes puzzled at the sight of a stooping peregrine, a dragonfly chasing a gnat, the humble housefly deftly outmaneuvering the descending swat or 400 tons of jumbo jet ponderously rising into the air to weave its way freely through "space," apparently defying the laws of nature. Even some professional pilots admit to feeling mildly surprised when their machines eventually leave the runway.

In addition to its wonder and technical fascination, flight is also one of the most beautiful and exhilarating activities to observe in insects and birds and to experience as a pilot or a passenger in an aircraft. This book seeks to explain how animals and man have evolved wings and mastered the associated physical problems and will attempt to reveal some of the magic of winged flight; the approach taken is a holistic rather than an anthropocentric one, embracing all forms of winged flight and placing nature firmly at the center.

With the exception of balloons, rockets and arguably helicopters, almost all man's flying machines— and certainly all flying creatures— depend on wings to support them in the air, so this book concentrates on winged flight. Unfortunately, the matter of aerodynamics is an unfamiliar and difficult branch of physics that embodies concepts and

Northern flying squirrel (Glaucomys sabrinus).

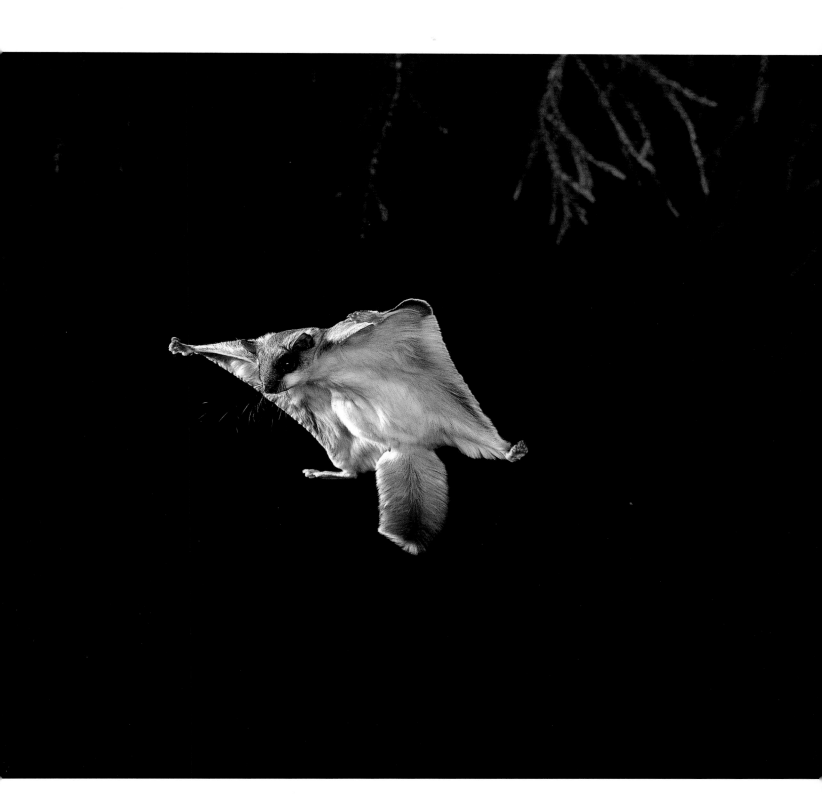

forces created by the movement of an invisible and intangible medium, yet it is impossible to explain how animals or aircraft travel through the air without describing the basics of the subject.

The main purpose of the first chapter, then, is to explore the physical principles of flight, but in the interests of simplicity, the more easily understood analogies of fixed-wing aircraft rather than of animal flight are generally used. This chapter, as well as all subsequent ones, is illustrated with diagrams and photographs that will clarify some of the more technical points in the text and aid in the understanding of the function, evolution and fascination of flight. Finally, this chapter touches on the intriguing recent discoveries about animal flight, for which standard aerodynamic theory cannot adequately account. Not until research supported by the evidence of high-speed photography revealed that insects and birds make use of nonsteady airflow did the differences between fixed-wing and flapping flight begin to be understood.

The second chapter is about the pioneers of all flight: insects. In it, we discuss the various ways in which insects could have evolved wings to become the first creatures to fly, 350 million years ago. Later, there is a description of the hitherto unknown and intriguing mechanisms that insects have evolved to obtain extra lift from the forces generated by air. Much of the information relating to this new but all-important aspect of animal flight was kindly provided by Charles Ellington at Cambridge, who has been studying insect flight in the laboratory for over 20 years. The third chapter, "The Feathered Wing," deals entirely with bird flight, starting with the evolution of birds from primitive feathered dinosaurs and going on to describe the anatomy and physiology of the bird in relation to its flight, whether gliding, soaring, flapping or hovering.

The last two chapters are devoted to the flight of man. In just a few generations, manned flight has come all the way from the floundering efforts of the medieval tower jumpers to the epic achievement of the first sustained, powered and controlled flight by the Wright brothers in 1903, which translated the mythical figures of Daedalus and Icarus into real life. The final chapter goes on to explain how a conventional modern airplane flies and, by contrast with the microaerodynamics of insect flight, includes man's latest technical achievements of supersonic flight before briefly reflecting upon the future.

The book would be seriously deficient without special mention of that most magnificent of flying machines, the Concorde, an aircraft which was a twinkle in the designer's eye as long ago as the late 1950s and yet will still be flying well into the 21st century.

As the story finally unfolds, we discover how the laws of aerodynamics are not as predictable as we might have first imagined but appear to change several times—from the flow patterns created by small, slow creatures with oscillating wings to conventional, transonic, supersonic and hypersonic regions of flight. Having come full circle, as it were, we finish with hang gliders and microlights, as once again enthusiasts revert to

Geological time	Millions of years	Life	Flight events
TODAY	—	Man	
	—		
	—		
TERTIARY	—		
	—	Modern mammals	
	— 50		
	—	Modern bird orders Modern insect orders	
	—		
	—	Bats	
CRETACEOUS	— 100		
	—		
	—		
	—		
	—	Early flowering plants	
	— 150	Early birds	
	—		
JURASSIC	—		
	—		
	—	Pterosaur	
	— 200	Early mammals	
	—		
	—		
TRIASSIC	—	Early dinosaurs	
	—		
	— 250		
	—	Mammal-like reptiles	
PERMIAN	—		
	—		
	—		
	— 300		
	—		
	—		
CARBONIFEROUS	—	First winged insects	
	—	Early reptiles	
	— 350		
	—		
	—		
DEVONIAN	—		
	—		
	— 400	Early insects	

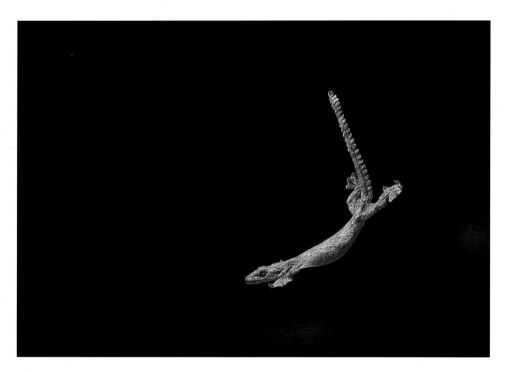

ABOVE: *Flying gecko* (Ptychozoon kuhli). FACING PAGE: *Blue-and-yellow macaw* (Ara ararauna).

leaping off hillsides and cliffs on flimsy wings. What is so difficult to imagine is the overwhelming timescale involved in all of this: Whereas insects gained their wings 350 million years ago, we learned the trick only a hundred years back—equivalent to a mere eight feet in the 6,000-mile journey from London to Johannesburg.

As this book is concerned with active sustained flight, there are no chapters devoted to "flying" creatures that are capable of remaining airborne for only relatively short glides. For instance, there is a flying frog that can extend its leap by spreading a membrane between its toes, as well as a lizard, a gecko, some squirrels and lemurs which glide on "wings" of skin stretched between their ribs or legs, but none of them have any means of airborne propulsion and so cannot improve on a simple glide.

The only animal that attempts powered flight is the hatchetfish of South America. After gathering

speed at the water's surface, it kicks itself out of the water with its tail and uses unusually large muscles for flapping its pectoral fins. However, its flight carries the creature for only a few feet, and one is left wondering whether a good leap would be a more effective way of escaping a pursuer. It will be interesting to see how the hatchetfish's descendants have fared in 100 million years or so.

As far as we know, the only groups of animals that have ever truly mastered powered flight are insects, pterosaurs, birds and bats, but insects alone have evolved new appendages as wings. Yet there is little direct evidence as to their origin. Indeed, apart from the *Archaeopteryx* and some later birdlike fossils unearthed in China, there is scant fossil evidence to explain how any creature acquired the power of flight, although present-day gliding animals do provide clues as to the first stages in the evolution of vertebrate flight. Moreover, apart from the hatchetfish, there are no feebly flapping animals alive today, so we can merely speculate as to how structures may have evolved into flapping wings.

The remaining animals that gained the power of true flight are pterosaurs and bats, but for reasons which are still uncertain, pterosaurs are not around today. Bats, on the other hand, are very much with us; they are impressive on the wing and are, not surprisingly, among the most successful group of mammals on Earth. In spite of major anatomical differences in their skeletons and wings, however, their flight is broadly similar to that of birds, so they do not merit a separate chapter.

CHAPTER 1

The Fundamentals

To most of us, flight is a miracle. To primitive man, it certainly must have seemed one. Flight still remains mysterious, even though the physical laws that govern it are now explained by the science of aerodynamics.

In his early attempts to fly, man tried to imitate the flapping action of a bird's wings, but because he knew nothing about the properties of air or the forces that such wings harnessed, his efforts failed. Centuries later, after experimenting with fixed wings, man started to raise himself from the ground, and at last, he began to understand the nature of the medium into which he was venturing. He learned, as the insect and bird had learned millions of years before him, by actual contact with the air's forces.

Yet the ingenious methods he evolved appeared to have little in common with those of the bird whose performance he had tried to emulate. Strangely enough, in the light of contemporary knowledge, the bird and the airplane seem to function in much the same way.

Birds and insects have been on the wing for hundreds of millions of years, but their performance and their "knowledge" of the medium in which they fly are innate and instinctive.

For man, however, flying is a matter of extreme effort and advanced technology. To appreciate the intrinsic qualities of flight in all its biological and mechanical diversity, it is essential to have some understanding of the principles of aerodynamics, which, in broad terms, is concerned with the motion of air, its displacement, its speed and the acceleration of its particles. After a brief summary of the history of the science of aerodynamics, this chapter explains the fundamentals of flight, but in the interest of simplicity, a basic airplane is generally used to illustrate the points.

A HISTORICAL REVIEW

The first serious attempt in Western history to analyze flight was made by Aristotle around 350 B.C. He maintained that an object, such as an arrow, could continue to move through the air only as long as a force was applied to it and that once this force was withdrawn, the object would stop. Furthermore, since he could not conceive of how a force could be conveyed from a distance, he claimed that the transmission of the object required a force in physical contact with it. Aristotle thereby concluded that an object could not be transmitted in a vacuum and asserted that the air, or atmosphere, actually perpetuated the flight. An arrow, he argued, traveled through the air by being pushed along by the air rushing in to fill the vacuum behind it. He presumed that the air sustained the flight of the arrow rather than retarded it. The Greek philosopher's

Whether upside down or right side up, as with the biplanes flying in an aerial display, above, the requirement of flight is a controllable force directed upward to overcome the force of gravity. FACING PAGE: *Noctuid moth.*

Figure 1. Archimedes' principle in action. A balloon will float upward if its total weight, with passengers, is less than the weight of the air that it displaces.

views on physics were not challenged until the 15th century, when Leonardo da Vinci (1452-1519) discarded Aristotle's central thesis.

Leonardo's assumption that the air was a resisting, rather than a sustaining, medium was the first approximate explanation of the phenomenon of flight. It is on this basic concept that the whole science of aerodynamics rests. However, his analysis of bird flight was incorrect: He supposed that the motion of the bird's flapping wings caused the air beneath to "condense" and to behave like a rigid body upon which the bird was supported. Using the same hypothesis, he explained that gliding flight depended on the relative motion between the wing and the air so that in a suitably strong wind, the bird could be supported without beating its wings. A kind of condensation process resembling Leonardo's notion does, in fact, occur when wings are moved through the air, but only at extremely high speeds approximating the speed of sound, when the density of the air changes significantly due to its compressibility (see Chapter 5).

One hundred years later, Galileo Galilei's (1564-1642) suggestion that motion could exist on its own dealt the death blow to Aristotle's theory. Like Leonardo, Galileo recognized the action of air in resisting motion and tried to define the precise way in which resistance changes with velocity. But a true scientific analysis of flight did not fall within man's grasp until Sir Isaac Newton (1642-1727) laid down the laws of motion and universal gravitation, which allowed his successors to study the properties of liquids and gases in a scientific

manner. From that moment, the secrets of the miracle of flight were gradually revealed to the world.

The chief requirement for flight is a controllable force directed upward to overcome gravity. A balloon, a bullet and an airplane are all capable of flying, in the broad sense of the word, but the reasons that each is supported in the air are entirely different. In every case, the force of gravity, which urges their return to Earth, is overcome. When a balloon floats in the air, it does so for the same reason that a cork floats in water: Any object suspended in a liquid or a gas experiences an upward thrust (or loss of weight) equal to the weight of the medium it displaces. This was first discovered by Archimedes in the third century B.C. For example, a balloon that occupies 1,000 cubic feet of air is subjected to an upward thrust of about 80 pounds (which is equal to the weight of 1,000 cubic feet of air). If the total weight of the balloon with its contents is less than 80 pounds, it floats upward. Whether it floats or sinks will depend on the difference between the volume of air it displaces and its own total weight (see Figure 1).

The problem with a balloon is that it cannot be fully controlled, and as far as direction is concerned, it is entirely at the mercy of the wind. True flight requires that ascent, descent and movement in any desired direction should be possible irrespective of the wind. An airship, which is really a rigid balloon with a means of propulsion, is capable of making headway in light winds but, nevertheless, is inefficient and ponderous.

On the other hand, neither a bullet nor a rocket is supported by

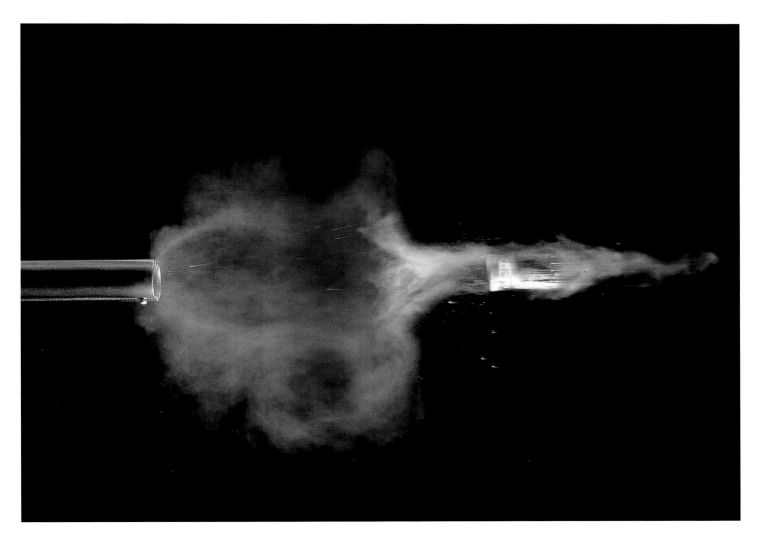

the air. They rely on their momentum and, in the case of a burning rocket, thrust. Together with all flying objects, they obey Newton's first law of motion, which states that a body stays at rest or continues to move in a straight line and at a steady speed unless another force acts upon it to effect a change. A bullet gains its initial thrust in the chamber of a rifle by means of an explosive charge, whereby it is rapidly accelerated up to three times the speed of sound before it leaves the muzzle. In the vacuum of outer space, the bullet would continue moving in a straight line and at a constant speed almost indefinitely, but in our atmosphere,

where other forces act upon it, its trajectory is shaped by gravity, which steadily pulls the bullet downward, and its speed is slowed by air resistance, or drag. The essential difference between a projectile and other flying objects is that a projectile does not rely on an atmosphere for its support; in fact, it becomes far more efficient in a vacuum, where its momentum is not dissipated by the air.

So, in the strict sense of the word "fly," only aircraft, birds, insects and bats fly, because winged flight depends on aerodynamic effects, whereby lift is derived from the movement of air flowing around the outer surfaces of a body. In order to

Thrust drives a bullet through the barrel of a gun, but this soon dissipates when it leaves the muzzle. The bullet— or, in this case, a 12-bore discharge— now depends on momentum to carry on its trajectory until friction and gravity return it to Earth.

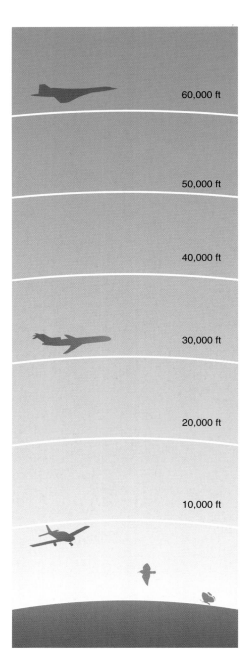

Figure 2. The flight of insects, birds and aircraft can take place only in the comparatively shallow belt of air surrounding Earth—the atmosphere.

understand these effects, we must first consider the properties and behavior of air.

AIR & ITS MOVEMENT

Both air and water are fluids, but air is a gas and water a liquid. One of the differences between a gas and a liquid is that a liquid is almost incompressible, while a gas compresses more easily. However, when considering low-speed flight, the matter of compressibility can, for all practical purposes, be ignored, as it does not assume any real importance until the object is approaching the speed of sound (about 750 miles per hour at sea level). At this speed, air becomes compressed, or to put it another way, it changes density. The reason this speed is significant is that sound is transmitted through the air in the form of waves which successively compress first one part of the air, then the next.

When an object is flying at a velocity less than the speed of sound, pressure waves precede the object and "warn" the air in front that the object is on its way. This enables the air to move out of the path of the object and to pass to one side or the other. But when the object travels at more than the speed of sound, things are very different—the warning wave cannot travel fast enough to get in front of the object. Now, instead of dividing and passing smoothly around the object, the air strikes the object with a shock and becomes compressed.

Needless to say, insects and birds do not have to concern themselves with supersonic flight and shock waves, in spite of the often-quoted but unaccountably inaccurate observation that the deer botfly can travel at 800 miles per hour! Nature has not found it necessary to travel at such speeds.

Movement of air provides the momentum that enables wings to overcome the force of gravity. We live at the bottom of an enormous ocean of air, which we tend not to notice until it starts to move. The faster it moves, the more we notice it. Although we have been using the movement of air to drive windmills and sail ships for thousands of years, we do not appreciate its extraordinary power until it reaches very high speeds, as in the case of the terrible destruction wrought by a hurricane or the remarkable aerodynamic forces necessary to support a 500-ton jumbo jet crammed with 400 passengers.

Winged flight can function only in our atmosphere, which is a comparatively shallow belt of air surrounding Earth. As altitude increases, the atmosphere becomes progressively thinner; the maximum density occurs at ground level (see Figure 2).

The air is held near the Earth's surface by the force of gravity or, in other words, by its own weight, which at sea level is about 0.077 pound per cubic foot. Because of this weight, the air exerts a pressure on Earth of about 15 pounds per square inch at sea level. The farther we move from the Earth's surface, the lower the pressure, so at an altitude of 20,000 feet, it falls below 7 pounds per square inch. Since air is a fluid, its pressure is transmitted equally in all directions: upwards, sideways and downwards, including any space to which it can gain entry. But if, for some reason, the air pressure on one side of an object is reduced, either the object or the air will have a tendency to be "sucked" toward the direction of lower pressure (or actually forced, due to the difference in pressure on each side). This is the principle behind all winged flight, whether animate or man-made.

RESISTANCE, LIFT & DRAG

Moving air has very different properties from static air, for as soon as an object and air start moving in relation to each other, another force begins to exert its influence. This force is so familiar that it is accepted without a second thought, and yet all winged flight, whether natural or mechanical, depends on it.

The force in question is resistance or, to be more precise, the air's resistance to motion. It is curious that an understanding of its origin and nature came much later in man's history than an understanding of the distances and movements of the stars and planets. The effect of the air's resistance to an object can be felt in two ways: It can appear as drag, acting in the direction opposing that of the motion, and as lift, acting perpendicularly to the direction of motion. When a body is flying at a constant speed in straight and level flight, its thrust balances the drag, while its lift opposes the force of gravity by balancing its weight. In aerodynamics, it makes no difference whether the body is at rest with the air flowing past it or the body is moving through still air—it is the relative motion that counts.

The forces generated in a stream of air depend on several factors: air density, airspeed, the area of the surface meeting the air, the shape of the body and the angle at which the body meets the air. The first factor is straightforward. If the density of

the air is doubled, then twice the weight of air will flow over the surface, thus doubling the forces generated. But with airspeed, the forces increase to the square of the velocity; that is, if the speed is doubled, then the aerodynamic forces increase four times. (This simple rule, however, does not apply at very high airspeeds.) If the area of the surface exposed to the airstream is doubled, then the aerodynamic forces produced will also double. Finally, the relationship between the shape of the body and the angle at which it meets the airstream is much more complex and requires more detailed explanation.

Unless there is some obstruction, a stream of air will take the shortest route from one point to another of lower pressure. Any obstacle in the airstream will force the air to deviate from its normally straight path, producing a reaction on the obstacle in the form of resistance. The greater the deviation inflicted upon the air, the greater the resistance, so naturally, the shape of the body in relation to the airflow has a significant effect. The best way to illustrate this is by drawing imaginary lines—streamlines—to indicate the direction of airflow at any given point.

But the flow picture revealed by a pattern of streamlines is more than just a chart of flow direction. It also depicts something very important—the relative velocity of the airflow. Where the streamlines are close together, the velocity is high, and where they are separated, the airflow moves more slowly.

A thin plate held with its edge toward the airstream offers minimum resistance, because it causes minimum deviation in the path of the air, which follows a steady,

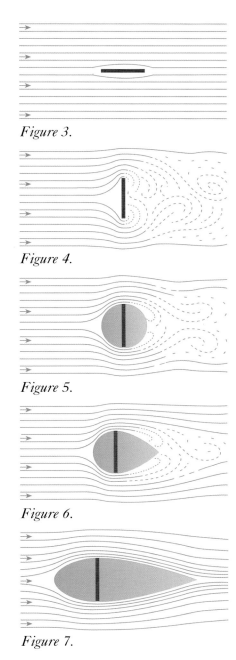

Figure 3.

Figure 4.

Figure 5.

Figure 6.

Figure 7.

Figures 3-7: Streamlining.
In Figure 3, a flat plate is edge-on to the airstream, offering minimum resistance. Maximum resistance occurs when the plate is held at right angles (Figure 4). Figures 5-7 show how "filling in" the spaces in front and behind the flat plate progressively improves streamlining.

Although the day-flying Pericopid moth from the Venezuelan cloud forest is flying forward, the wings are in the middle of their upstroke and so have a smaller angle of attack than might first appear.

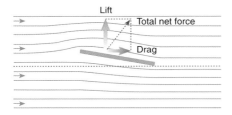

Figure 8. Angle of attack and forces acting on a flat plate.

smooth route (see Figure 3). But the same plate held at right angles to the airflow increases the air resistance, or drag, several hundred times (see Figure 4). In this position, the air in front must change its direction drastically, while the airflow behind the plate becomes broken up into eddies, producing turbulence. The greater the turbulence, the greater the drag.

Drag can be dramatically reduced by streamlining the object so that the turbulent spaces behind are filled in and the front areas are rounded or tapered (see Figures 5-7). By

Figure 9. Balance of forces on body flying at constant height and speed.

streamlining the flat plate, its drag can be reduced by as much as 95 percent. In fact, filling in the turbulent area behind the object improves streamlining more effectively than does altering the front. As drag increases with the square of the airspeed, it follows that the higher the speed required, the more streamlining is necessary. The ratio of the length to the breadth of a streamlined body is known as its fineness ratio; high fineness ratios improve streamlining efficiency, as is reflected in the extremely thin wings of supersonic aircraft. That part of the drag which is due to the shape of the object and which can be reduced by streamlining is known as form drag.

The nature of the body's surface also affects drag. It is easy to see that a rough surface will generate more skin friction as a result of the air flowing over it than will a smooth one. With high-speed flight, the effects of skin friction are considerable, so a smooth finish becomes an essential element in the design of fast aircraft. If an aircraft is to meet with minimum air resistance, then the first consideration is to reduce form drag by designing the most suitable shape for the rear and front and by eliminating projections and sharp corners. After this, attention can be given to reducing skin friction. A good definition of a streamlined body is one in which the form drag is less than the skin friction (see also "The Boundary Layer" on page 28).

The angle at which a body meets the airstream, or angle of attack, is also a crucial factor. It is important to realize, however, that the angle of attack has nothing to do with the horizontal or with the position of Earth; it is simply the angle at which the object meets the flow of air.

Referring back to the thin plate, we know that there is minimum air resistance when the plate is held with its edge to the airstream and maximum drag when it is held at a right angle, but what happens when the leading edge is inclined at a slight angle (see Figure 8)? The air pressure is now greater underneath the plate than on the top surface, producing a lifting force that acts at right angles to the undisturbed airstream. But whenever lift is generated, drag is also created, and this acts in the opposite direction to the motion of the object. Whereas lift acts at right angles to the airstream, drag acts parallel to it. The net result is that the total force on the plate will be acting backwards as well as upwards.

Naturally, it is lift that makes heavier-than-air flight possible, while drag tends to prevent it. Both are really part of the same force, but as they have very different effects, it is important to distinguish between them. If an aircraft or a bird is to fly at a constant height and speed, the lift acting on it must equal its weight, while the drag must be equal to the driving force, or thrust. So the greater the lift, the greater the drag and the greater the thrust required to balance it (see Figure 9).

An ideal airplane or bird would be all wing, free from such projections and protuberances as engines, tails, fuselages, heads, wheels, legs and any other extra parts that produce drag without directly contributing to lift. Drag in this form is appropriately called parasitic, and in an aircraft, everything is done to keep parasitic drag to a minimum. But the drag that is actively responsible for producing lift, the induced drag, cannot be eliminated, even

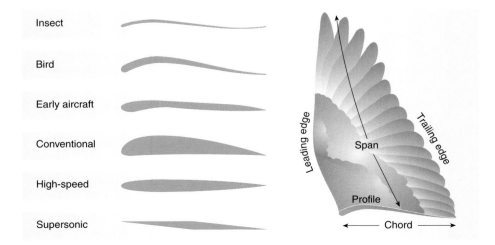

Figure 10A, far left. Airfoil sections.
Figure 10B, left. Parts of a wing.

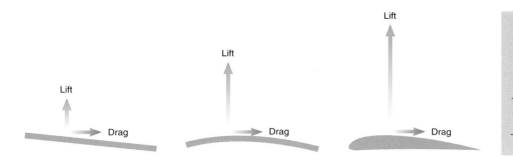

COMPRESSIBILITY
For all practical purposes in aerodynamics, air becomes compressible only as it approaches the speed of sound (see supersonic flight, page 156).

Figure 11. Ratios of lift to drag of flat plate, cambered plane and airfoil.

Bernoulli's principle is demonstrated by blowing between two sheets of paper.

in an ideal aircraft, since it is a necessary adjunct of lift. More will be said about this later.

AIRFOILS & WINGS

The whole principle of wings, whether they be of bird, insect or airplane, is that they are designed in such a way as to push the air downward and, in so doing, gain an upward reaction. It is rather like walking up a sandy slope—every step upward pushes the sand downward. Both examples obey one of Newton's laws: "To every action, there is an equal and opposite reaction." The downward flow of air created by wings is called downwash. The more air deflected downward by a wing, the greater the lift; on the other hand, the more air disturbance produced, the greater the drag.

Although a flat object can function well as a simple wing at low Reynolds numbers, as demonstrated by the thin balsa-wood wings of a simple model glider, a wing's lifting and general aerody-

namic characteristics can be dramatically improved at high Reynolds numbers if it is shaped and streamlined in a certain way. Wings which are designed and angled to the airstream in such a way that the maximum downwash and lift are obtained with the minimum of turbulence and drag are known as airfoils (see Figure 10A). The gentle curvature of the airfoil entices the air to flow smoothly over the upper surface in a downward direction without breaking away and forming eddies. This is how the conventional airfoil scores over the flat or cambered plane (see Figure 11).

The peculiar properties of airfoils rely on a natural law concerning the relationship between the speed at which a fluid moves and the pressure it creates. Formulated by Swiss mathematician Daniel Bernoulli (1700-1782) some 200 years ago and known as Bernoulli's principle, this law affirmed that the greater the velocity of a fluid, the less pressure it exerts. The principle can be demonstrated quite simply by holding

Figure 15A. Movement of center of pressure with angle of attack.

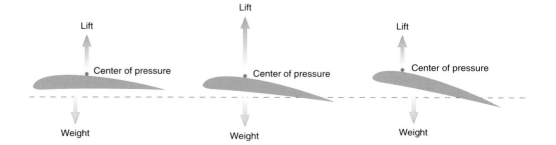

two pieces of angled paper an inch or two apart and blowing between them. Instead of opening up, the two sheets close together.

Because air behaves like an incompressible fluid at normal speeds, it has to accelerate to pass through the restriction. It reduces the pressure inside, so the two sheets are forced together by the relatively higher pressure on the outside surfaces. Another experiment that demonstrates Bernoulli's principle is to hold a tablespoon downwards between finger and thumb and direct a jet of tap water onto its convex surface. Instead of pushing the spoon away, the water hugs the curvature, pulling the spoon into the jet.

Both of these experiments illustrate the drop in pressure that occurs when the speed of air or water increases. The shape of a spoon is markedly similar to the wing of a bird or an airplane, and in fact, all three behave as airfoils. When a stream of air passes around an airfoil, the air flowing over the more convex upper surface has a greater velocity and, following Bernoulli's principle, a lower pressure than those on the undersurface. The difference in pressure between the two creates the lift (see Figure 12). Unlike a flat plate, an airfoil with a humped upper surface will produce lift even when the angle of attack is zero degrees. Further lift is gained by increasing the angle of attack so that the air meets the undersurface at a steeper angle (see Figure 13).

The lift derived by a wing is the sum total of all the pressures acting upon its surface. It is by no means equally distributed—normally, about two-thirds of the lift is due to the decrease in pressure on top, while one-third is due to the increase in pressure underneath (see Figure 14).

For the sake of convenience, however, the total aerodynamic force can be considered to act from a point known as the center of pressure, whose position alters with the angle of attack (see Figure 15A).

For any given wing section, lift, drag and center of pressure are largely controlled by the angle of attack. As the angle of attack increases, so does lift, while the center of pressure moves forward. Un-

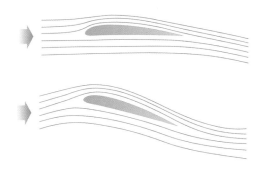

Figure 12, TOP.
Airflow around an airfoil.
Figure 13, BOTTOM.
Effect of increasing the angle of attack.

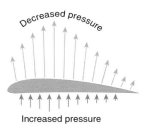

Figure 14. Pressure distribution over airfoil surface.

LIFT-TO-DRAG RATIOS

The forces of lift and drag over an airfoil are best demonstrated by a graph of lift-to-drag ratios. As shown on the bottom axis, lift and drag depend on the angle of attack. At zero degrees, both lift and drag are low, but as the angle of attack rises, lift increases faster than drag, up to a maximum of around four degrees—the best lift-to-drag ratio. Thereafter, drag increases more rapidly, and at about 15 degrees, the wing stalls due to the turbulent airflow on its upper surface. Now lift has disappeared, but drag continues to build up.

Figure 15B.

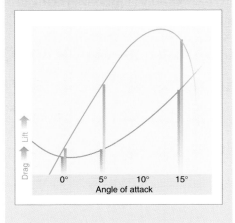

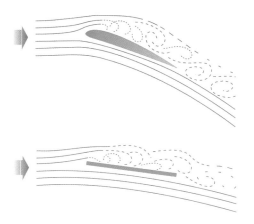

Figure 16. Stalling angles of flat plate and airfoil.

Figure 17. Air "leaking" from underneath a wing to the top surface.

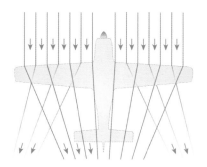

Figure 18. The deflection of the airflow above and below the surface of a wing causes trailing vortices.

fortunately, drag increases at the same time, slowly at first, but at a certain point, it starts to increase more rapidly than lift. Any particular wing, therefore, has an optimum angle of attack, providing the best ratio of lift to drag. Most aircraft are designed to cruise at this angle, because they consume the least fuel. Typical aircraft airfoils have the most efficient lift-to-drag ratio at about 3½ to 4 degrees.

As the angle of attack is increased, the gain in lift cannot be indefinite, because at a certain stage, the air ceases to flow smoothly over the wing's upper surface. Although the undersurface may still be providing lift, the air on the upper surface no longer flows smoothly around the hump but breaks away into turbulent eddies. At this point, the wing loses its lifting properties and is said to stall.

The angle at which stalling occurs is called the stalling angle (see Figure 16). With aircraft, the stalling angle is at about 15 degrees, which is much higher than with a flat plate; in some insects, it may be as high as 60 degrees. The suddenness with which this comes about in aircraft can be dangerous, particularly as the pilot sometimes has little direct control over the angle of attack.

It should be remembered that the angle of attack depends on the relative airflow to the wing; thus a stall can occur at any attitude. If an airplane is flown too slowly, for instance, it will sink and so meet a relative airflow from underneath, causing the stalling angle to be reached, even though the angle, or attitude, of the airplane may be horizontal to the ground. For this reason, it is more usual to talk about stalling speeds than stalling

angles. It is important not to confuse stalling with sinking; any aircraft or bird will sink when the total upward force is less than its weight, but stalling takes place when the angle of attack or the airspeed is such that the air suddenly separates from the upper surface of the wings and becomes turbulent. The more practical aspects of stalling are discussed in later chapters, together with the various devices used by birds and aircraft to reduce their stalling speeds.

As well as providing lift, the difference in pressure between the lower and upper wing surfaces is responsible for a related effect known as wingtip vortex. Explained simply, wingtip vortex is due to the tendency of air, or any fluid, to flow from high to low pressure. As there is no barrier at the wingtips separating the high- from the low-pressure areas, the air leaks around from underneath the wing to the top surface (see Figure 17), causing the air on the top surface to be deflected slightly inward and the air on the bottom surface to flow outward (see Figure 18). As a result, the streams meeting at the wing's trailing edges cross one another to form a series of small trailing vortices, which then join up into one large vortex at each wingtip (see Figure 19).

The energy that goes into the formation of these vortices appears as the previously mentioned induced drag. Wherever there is lift, there must also be induced drag, as it is caused by the wings in the generation of lift. Unlike other types of drag, induced drag does not increase with the square of the airspeed. On the contrary, induced drag is at its greatest when maximum lift is being achieved at the

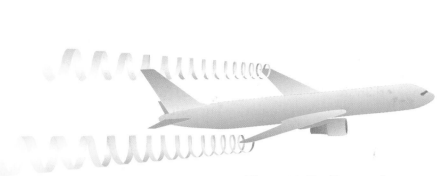

Figure 19. Trailing vortices.

Figure 20A. The efficient high-aspect-ratio wings of the sailplane and albatross are designed to fly with a minimum expenditure of power, whereas the low-aspect-ratio wings of many jet fighters and birds such as the wren may be less energy-efficient but allow a degree of maneuverability.

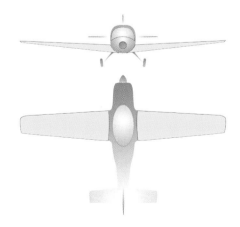

Figure 20B. Wing taper in elevation and plan helps to reduce induced drag.

WINGTIP VORTICES

Wingtip vortices possess enormous energy and are considered to have semi-infinite life in calm conditions. They have been responsible for several accidents caused by a loss of control of the aircraft. Some aircraft generate far more wake turbulence, as it is known, than others—that generated by the Concorde is particularly severe. When flying behind one another, all aircraft should take account of this phenomenon, especially at high angles of attack during landing, when the vortices are most powerful.

lowest airspeed—just before stalling (at the maximum lift coefficient; see Figure 15B). This is because to maintain lift, the angle of attack must be increased as the airspeed is reduced, which induces a greater difference in pressure between the upper and lower surfaces of the wing. Consequently, there are more violent wingtip vortices and correspondingly more induced drag.

One way of keeping induced drag low is to extend the wings so that the tips make up a relatively small proportion of the wing. Aircraft and birds with long, narrow wings have what is called a high-aspect ratio (ratio of span to chord), while those with short, stubby wings have a low-aspect ratio (see Figure 20A). Another way to minimize induced drag is by incorporating wing taper; again, this reduces the relative proportion of the wingtip (see Figure 20B). Tapering in both depth of airfoil section (elevation) and plan form has advantages from an aerodynamic and a structural point of view and is a feature we were slow to learn from nature. Yet we have only to look at the structure of a bird's wing to see this.

The main advantage of high-aspect ratios and wing tapering is that by cutting down the induced drag, they provide maximum lift with minimum consumption of power. That is why gulls, gliders and airplanes designed for long-distance flying have high-aspect ratios—long, narrow wings are simply more efficient, particularly at low speeds. The advantage of a high-aspect ratio is taken to the extreme when several aircraft or birds fly in formation, with wingtips just behind adjacent wingtips. This demonstrates the theory of circulation and wingtip vortices; the combination is equivalent to an aircraft or a bird of several times the span and results in a significant improvement in efficiency and maximum range.

But, like most things, aircraft and

Figure 21.
Unstable and stable handcarts.

birds with high-aspect ratios have their snags: As well as being more difficult to maneuver, both in the air and on the ground, they tend to be heavier structurally, so a point is reached when the benefits of increased lift are cancelled out by the extra weight. This is one of the main reasons that man-powered aircraft have not been more successful. To minimize the induced drag, which at low airspeeds consumes a high proportion of the power, it is beneficial to have wings with high-aspect ratios, but the extra weight they carry imposes a practical limit on the wingspan.

Any means of reducing drag without sacrificing lift or, conversely, increasing lift without increasing drag must obviously improve the lift-to-drag ratio. An alternative way to minimize induced drag at low speeds is by having a large wing area to support the weight of any given body, or low wing-loading, in aeronautical language. This simply means that as the surface area of the wing is larger, each square foot has less weight to support. The chief disadvantage of low wing-loading is that the wings are relatively large, thereby creating more surface friction. As skin friction increases with the square of the airspeed, birds and aircraft that are efficient at high speeds, such as wrens and jet fight-

ers, have small wings and, hence, high wing-loading and are therefore better suited to sudden maneuvers. Sailplanes and vultures, on the other hand, which have large wings and correspondingly low wing-loading, are ideal for cross-country gliding flights, which consume minimum energy.

STABILITY

Flight cannot be achieved simply by attaching a power unit to a wing and taking off. Obviously, such a contrivance would lack stability and be hopelessly uncontrollable. Any vehicle, be it a car, a ship or an airplane, requires some degree of stability. A two-wheeled handcart is unable to remain balanced unless stabilized by a driver through a shaft (see Figure 21).

Likewise, a wing would be equally unstable unless balanced in some way. Imagine a wing flying along with its lift and weight perfectly in balance; the equilibrium would be upset by the slightest air disturbance. If, for instance, the front of the wing were lifted up by an air current, the angle of attack would increase (see Figure 22). Remember that the center of pressure shifts forward with an increasing angle of attack, so the two lines of force would become out of line with each other. This would produce an even greater angle of at-

Figure 22. Instability of a simple wing.

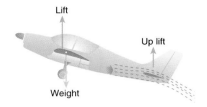

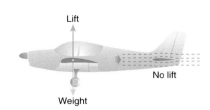

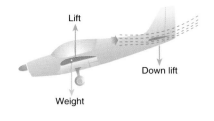

tack, resulting in a further forward movement of the center of pressure, inducing the wing to start revolving uncontrollably around its long axis. A practical demonstration of this can be seen by dropping a postcard from a height.

With aircraft, the matter can be rectified quite simply by adding a fuselage, which acts as a lever in the same way as does the shaft of the handcart. In place of the controlling hand, a tailplane, or stabilizer, is added. The function of the tailplane is to create a force in the appropriate direction to compensate for any out-of-balance effects, providing what is known as longitudinal stability, or stability in pitch. Normally, by virtue of its symmetrical airfoil section and zero angle of attack, the tailplane provides no lift. But as soon as a disturbance causes a nose-up attitude, the tail drops and a positive angle of attack is produced. In this way, lift is generated, pulling the tailplane upward, and the wings are brought back to their original position (see Figure 23A).

In addition to longitudinal stability, measures must also be taken to prevent airplanes from wandering from side to side. Directional stability, as it is called, is achieved by including a vertical surface—the fin —above the tailplane. The fin functions in the same way as the tailplane but corrects in a directional sense. It should also be borne in

ALTERNATIVES TO TAILPLANES

A few aircraft, such as the Concorde and some supersonic fighters and microlights, do not possess tailplanes but have either sweepback or reflex or both built into the wings (see Figures 23B and 23C). With sweepback, the area toward the wingtip well behind the center of gravity acts as a fixed surface that provides stability in the pitching plane.

mind that the sides of the fuselage have a larger area behind the center than in front, so as soon as the machine "weathercocks" to one side, the pressure differential assists the fin in rotating the machine back to its original path (see Figure 24).

Finally, some method must be found to maintain lateral stability so that the flying body remains on an even keel and is prevented from rolling uncontrollably from one side to the other. If an airplane begins to roll, forces come into play to oppose the roll, because the wing that is moving downward effectively strikes the air at a greater angle of attack than the upward-moving wing. As each wing is now subjected to a different amount of lift, there will be a tendency for level flight to be restored automatically. However, once the airplane has tilted into a roll, its weight and lift become laterally out of line with each other, resulting in

Figure 23A. Stabilizing function of tailplane.

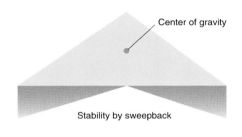

Figure 23B. Sweepback at wingtips.

Figure 23C. Reflex at trailing edge.

Figure 24. Directional stability provided by fin and side of fuselage.

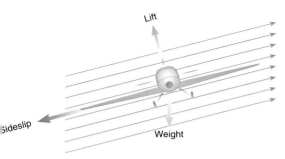

Figure 25. An aircraft in a sideslip.

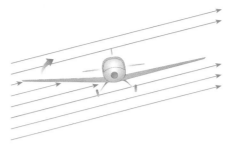

Figure 26. Use of dihedral for lateral stability.

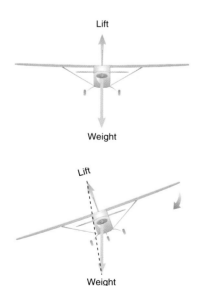

Figure 27. Lateral stability is achieved in a high-wing monoplane by keeping the center of gravity low.

the plane sideslipping toward its lower wing (see Figure 25).

Now, if the wings on each side are horizontally level with each other, the aircraft will have a tendency to carry on sideslipping, because the angle of attack on each span is the same. But by incorporating a dihedral so that the wings are arranged in a flat V, the angle of attack on each wing of a sideslipping aircraft is different (see Figure 26). The effect of the lateral airflow will be to raise the lower wing so that the aircraft assumes its former level position.

An alternative method of establishing lateral stability is achieved in a high-wing monoplane, where the wings are positioned on the top of the fuselage (see Figure 27). The center of gravity is kept low, so the machine behaves like a pendulum, counteracting the lateral tilt of the wings.

All-round inherent stability—the combination of all the stabilities mentioned—makes the pilot's task much easier, particularly in turbulent weather conditions, when the airplane is being tossed about in the sky. But as we shall see in Chapter 4, there was a time in the early days of flying when unstable airplanes were preferred to stable ones, which meant that the pilot was constantly fighting the controls to keep his machine in the air.

Nevertheless, stability should not be excessive, particularly with aircraft that must be highly maneuverable, because they would then have a tendency to resist any change in attitude or direction of flight. Insects and birds, on the other hand, depend for their survival on being able to maneuver rapidly and do not require the inherent stability of aircraft because they are continually monitoring and adjusting their flight by an instinctive reflex action acquired through millions of years of evolution.

THE BOUNDARY LAYER

Because of its internal friction, air (as well as any fluid) has a resistance to flowing smoothly. In other words, it has viscosity. For example, molasses is more viscous than water, and both exhibit different degrees of resistance to rate of change in shape. This is due to the tendency of one layer to "stick" to the adjacent layer, thus resisting relative movement between the two. Air is plainly much less viscous than molasses or water, but nevertheless, what viscosity it has is enough to cause skin friction. Ultimately, it is responsible for causing all turbulence, drag and even lift itself.

One of the greatest advances in the history of aerodynamics was made early in the 20th century by German mathematician Ludwig Prandtl (1875-1953). His discovery of the boundary layer introduced a new concept about the motion of fluids that has subsequently led to significant improvements in the design and performance of aircraft. If drag is to be kept low, we know it is essential that the air be induced to flow smoothly over all the surfaces with which it is in contact. But flow of air immediately adjacent to a surface has peculiar properties of its own—it doesn't actually flow over the surface at all.

The flow of air within the boundary layer is similar to the movement of the leaves of an open book lying flat on a table. When the upper corner is pushed at right angles to the spine, the pages slide over one another, the speed of the movement of the pages decreasing from the top downward. The book is said to be sheared, while the force that causes it is termed a shearing stress. The concept of the boundary layer assumes that no matter how smooth a surface may be, the molecules of air which are in actual contact with the surface remain motionless and do not move over it. Some distance above the surface, the air moves smoothly and at full speed, and sandwiched between this main flow and the stationary film is the boundary layer, where the velocity of the air molecules increases outward from the surface (see Figure 28).

This velocity gradient is confined to an extremely shallow layer of air that is normally about one-hundredth of an inch thick, and no matter how rapidly the object is moving—whether it is a bullet or a supersonic fighter—the relative velocity of the air at its surface is exactly zero. Since the airflow above the boundary layer is moving at full speed, it is not difficult to imagine the magnitude of the shearing stresses that must be created within. If the successive layers of molecules within the boundary layer slide smoothly over each other, then the flow is said to be laminar. But if the various layers are dragged along on top of one another, forming rolling eddies and vortices, then the flow is turbulent, resulting in increased skin friction and wasted energy. When the boundary layer breaks away from the surface, it gives rise to increased drag or stalling, but if the airflow within the boundary layer can be controlled by keeping the flow smooth and close to the surface, drag can be kept to the minimum and the stalling point delayed. There are a number of mechanisms employed both by flying creatures and by aircraft that increase lift and reduce stalling speed by maintaining a laminar flow in the boundary layer. These are described later.

When air flows over a wing, it usually does so in a laminar manner up to the thickest point, but after this, the flow tends to become turbulent within the boundary layer. Laminar-flow wings on aircraft are designed to preserve a smooth flow within the boundary layer as far back as possible (see Figure 29A). In this way, the transition from smooth to turbulent flow can be considerably delayed, reducing drag due to skin friction by up to 50 percent. The behavior of the boundary layer on the wings of insects likely plays a crucial role in the micro-aerodynamics of insect flight.

SCALE

Before we enter into the realm of microaerodynamics, a brief digression is necessary. Flying to humans seems a formidable task; to an elephant, it would appear even more so. But to a little spider, the undertaking would seem far less daunting; in fact, its problem is more about staying on the ground.

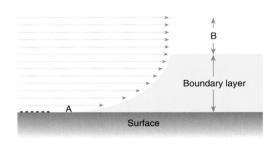

Figure 28. Effect of skin friction showing airflow decreasing speed toward surface. At A, the air is static; at B, it is moving at full speed.

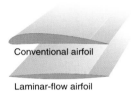

Figure 29A. A comparison between a conventional and a laminar-flow airfoil. In the laminar-flow airfoil, the maximum thickness of the profile is around the center, rather than toward the front.

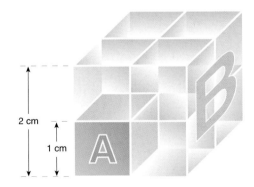

Figure 29B. Surface-volume relationship. Cube A is twice the linear size of Cube B and has four times the surface area and eight times the volume.

Conventional aerodynamic theory cannot really explain how any insect can fly, let alone a bumblebee. This photograph demonstrates the lifting power of an insect's wings as the creature raises its not inconsiderable bulk into the air. Note that the wings are twisted over 90 degrees on the upstroke.

for this follow some basic laws of physics. If the linear size of an object doubles, its area increases fourfold, while its volume (and weight) increases eightfold. Put another way, surface area increases with the square of the linear dimensions, and volume increases with the cube of the linear dimensions. The concept is known as the surface-volume relationship, and Figure 29B illustrates this relationship clearly (see also "Reynolds Numbers" on page 21).

This inescapable physical law goes to the very core of things both living and nonliving. It explains why small animals, such as bees and spiders, do not need surface-rich lungs, kidneys and a complex network of blood vessels, while large animals do. It also explains why small birds have disproportionately smaller wings than large birds. A large, heavy creature, such as a swan or a condor, needs relatively much more wing area to remain airborne than a small bird like a wren. An ostrich is so large that it has given up flying altogether—it would require such huge wings and strong, heavy muscles to power them that flight would become inefficient and impractical.

A human must employ wings the size of a hang glider to remain airborne (in a glide). But a small animal like a money spider does not need wings at all: It floats into the air on rising air currents and travels around the world as aerial plankton in the company of all sorts of other tiny creatures. Whether or not these aerial planktonic animals possess wings is irrelevant—they are not flying in the true sense, as they have little or no control over their ultimate destination.

Whereas a whirlwind could lift us into the air and a wind of extraterrestrial origin would be required to raise an elephant off its feet, a puff of wind is sufficient to send the spider floating over the trees.

It may seem obvious, but the smaller you are, the easier it is to become airborne. The reasons

NONSTEADY AIRFLOW

Aircraft and many flying creatures fly in what might be described as a "standard" way, using established aerodynamic principles. This, however, does not explain how insects, bats and small birds fly, how hover flies hover, how dragonflies fly backward or, indeed, how bumblebees and large beetles are able to leave the ground at all. And just consider the astounding aerobatic feats of the humble housefly. It can decelerate from rapid flight, turn within its own length, hover, fly upside down, loop, roll and land on the ceiling—all within a few milliseconds. The spectacular aerial performance of insects reigns supreme.

Although scale has a part to play here, conventional aerodynamics still does not provide many clues as to how insects gain sufficient lift to perform in this impressive way, but research at Cambridge University has revealed some fascinating new evidence that has helped to explain how the insect world has apparently been violating some of the laws of aerodynamics. It has been discovered that insects make use of nonsteady airflow to generate lift. The details of the actual mechanisms used are explained in Chapter 2, but the broad outline of the aerodynamic theory is discussed here.

Aircraft rely on airfoils that move through the air steadily. Any instances of nonsteady airflow around aircraft wings must be minimized, since they reduce the efficiency of flight and may even produce dangerous vibrations. By contrast, insects and birds depend on flapping wings that clearly produce varying degrees of nonsteady airflow. Thus nonsteady aerodynamics is an inherent feature of natural flapping flight,

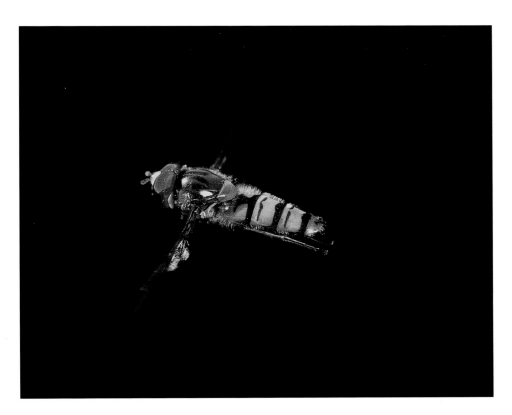

but its importance varies with the size and the airspeed of the animal.

Generally, the larger and faster the animal and the lower the frequency and amplitude of its wing beats, the less significant the effects of nonsteady airflow. It can be assumed, therefore, that when a bird is in fast forward flight, the flow of air around its wings is almost steady and its performance can be largely explained in terms of conventional low-speed aerodynamics.

The flight of the smaller and slower-moving creatures, however, is different. Their wing movements are far more extensive and have a faster frequency, resulting in aerodynamic forces that oscillate widely in the three dimensions of space and vary considerably in magnitude. The airflow becomes nonsteady, with the fluctuation increasing in intensity and significance as the airspeed diminishes, reaching a maximum in hovering flight.

Hover fly (Syrphus balteatus).

Long-eared bat hovering to pick an insect from the leaves.

bers, we saw that normal airfoil action functioned well at high and intermediate values and that it deteriorated significantly at lower values, when the viscous shearing forces and drag were large compared with the inertial forces and lift. In effect, this means that the smaller the animal (and the lower the Reynolds number), the larger the relative drag. How is it, then, that animals hover by flapping their wings? And how is it that very small insects fly at all? Before answering these questions, we must look at the nature of lift from another perspective.

If we could see a stream of air as it flowed past an airfoil, the air would actually appear to be circulating around the airfoil. Consider, for example, a horizontal revolving cylinder. If the cylinder were to be placed in an ideal fluid of zero viscosity, its rotation would not affect the fluid in any way, as there would be no friction or boundary layer produced. But when the cylinder is immersed in a real fluid that has viscosity, such as air, and spun around its axis in, say, a counterclockwise direction, the surrounding air is set in motion and rotates with the cylinder. This is called a bound vortex (see Figure 30).

Now, if a horizontal wind were to blow from right to left, the combination of the counterclockwise rotating air of the bound vortex and the horizontal airflow would cause an increase in the speed of the air above (compression of streamlines) and a decrease in the speed of the air below (see Figure 31).

The net result would be a decrease in pressure above and an increase below, due to Bernoulli's principle, together with an upwash in front and a downwash behind. In

Hovering is not confined to hummingbirds, hover flies and dragonflies. Most small birds can hover for short periods, particularly when courting or approaching their nests. Bats, too, are capable of hovering, and among the insects, the ability to hover is the rule rather than the exception. Although there is more than one type of hovering, it is usually a strenuous form of flight and makes peculiar aerodynamic demands upon the creature in question.

When discussing Reynolds num-

effect, the spinning cylinder would be subjected to lift in the same way as would an airfoil. This phenomenon is known as the Magnus effect. A similar kind of action can be seen when a sliced golf ball is sent sailing into the rough. As the ball is driven forward, it is set spinning by an oblique blow of the club, and the effect of viscosity induces the air to stick to the ball's surface and rotate with it (see Figure 32).

Where the rotary motion of the bound vortex and the relative airflow clash head-on at the right-hand side of the ball, the combined streams are slower than the streams on the left-hand side, which are moving in the same direction. As the streamlines are closer together on the left, the pressure differential induces the ball to be "sucked" to the left and, perhaps more often than not, to deviate from the path intended by the golfer. It is important to understand that without the bound vortex and its superimposition upon the airstream as the ball flies through the air, there would be no "lifting" force at all.

We can now study airfoil action in a new light. Think of it as a device that acts like the revolving cylinder for creating and maintaining circulation in the form of a bound vortex. If we were able to move with the undisturbed airstream and watch the airflow passing over a wing, it would seem as if the air were circulating (see photo on page 35).

The air in front moves upward and over the top of the airfoil, flowing faster than the main airstream, while the air underneath flows slower than the main airstream. Relatively speaking, the air appears to move in a circle. The idea of rel-

ative airflow around an airfoil is called circulation, but the particles of air do not actually circulate around the wing. This circulation, as we know, takes place outside the thin boundary layer, although the boundary layer initiates the flow.

Provided circulation is established, a wing will experience lift as soon as it is exposed to a wind, but a knowledge of the way in which the circulation is initially set up is important in understanding insect flight. What happens when a wing starts to move through the air from rest? If you hold a piece of inclined cardboard in smoke and move it from rest, you will see an eddy shed from its trailing edge.

Called a starting vortex, this is created because of the large viscous shearing forces close to the trailing edge of the wing and is generated every time a wing starts its movement. One of the rules of aerodynamics is that a vortex cannot be created without the production of a countervortex of equal strength circulating in the opposite direction (see Figure 34).

A moisture-laden golf ball is set spinning as it leaves the club. The dimples on the ball help to form a turbulent boundary layer, thus increasing air circulation around the ball.

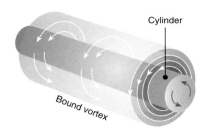

Figure 30.
Formation of bound vortex around cylinder revolving in still air.

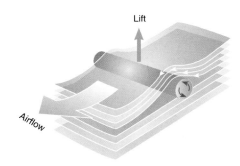

Figure 31.
Lift generated by a cylinder revolving in horizontally moving air.

Relative airflow

Figure 32. Sliced golf ball, with a counterclockwise spin, is subjected to a lifting force to the right (in diagram).

In this case, the countervortex is, in fact, the bound vortex, which is responsible for the circulation and production of lift, and it owes its continuing existence to the shearing forces over the surfaces of the wings. We now have to consider two opposing currents of circulating air at the start of the wing movement: the starting vortex and the bound vortex. Since they interact destructively in inverse proportion to their distance apart, the net lift around the airfoil at the start of the movement is minute.

But once the starting vortex has been left well behind (after the wing has moved about three chord lengths away from its starting point), then the smooth characteristics and lift of steady airflow are established. The interaction that delays the creation of lift is called the Wagner effect and constitutes an unavoidable nonsteady phase in the action of normal airfoils.

Let us now consider the effect on flapping wings. During hovering, in particular, the repeated stopping and starting at the end of each wing stroke would appear to hamper, rather than assist, the production of lift.

This becomes especially acute with small insects at low Reynolds numbers, when the creation of circulation becomes difficult and, furthermore, the vortices tend to die out rapidly. Another apparent disadvantage of the flapping wing is that when it comes to rest at the end of each stroke and the lift is reduced, the bound vortex has to be shed, becoming a "free" vortex, which, in hovering flight, will interact with the new starting and bound vortices that initiate the return stroke. Calculations made of the minimum lift

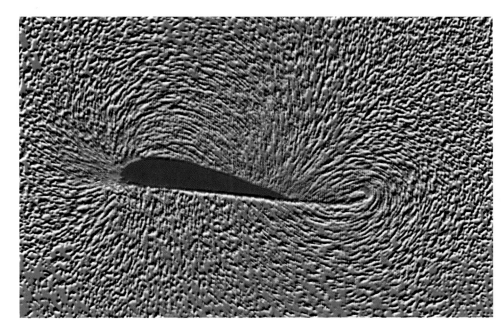

Circulation of liquid around an airfoil—the photograph was taken with the camera moving with the flow of liquid. The starting vortex is visible at the trailing edge (see also Figure 33).

coefficients necessary to sustain flight at low Reynolds numbers show them to be far in excess of what is possible in a steady-flow situation.

Clearly, then, standard aerodynamics has failed to furnish the answer to how insects and birds overcome these apparently insuperable handicaps. But thanks to the pioneering work of Charles Ellington and the late Torkel Weis-Fogh at Cambridge, as well as the evidence furnished by high-speed photography, the mystery today is nearer solution. We can now conclude that insects, birds and bats turn nonsteady airflow to their advantage by employing mechanisms which can swiftly establish air circulation around their wings, entailing the rapid formation and shedding of vortices. By so doing, they are able to generate far more lift than would be possible in steady airflow. This discovery has cast an entirely new light on the problems that have long perplexed observers of bird and insect flight and will be discussed in greater detail in the next chapter.

Figure 33. Starting vortex on the trailing edge.

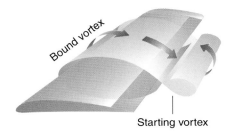

Figure 34. Starting vortex and the production of its countervortex, the bound vortex.

CHAPTER 2

The First to Fly

The sight and sound of wings are so much a part of our everyday experience that we tend to take for granted anything in the air, from dancing gnats to supersonic fighters. Indeed, it is difficult to imagine looking up into a sky devoid of insects, birds and airplanes. Yet up to a time unimaginably distant, the sky was, in fact, lifeless and silent. Then, from the dense undergrowth of steamy swamps and the edges of lagoons and lapping seas, anonymous insects began their first tentative flights—flights that were even more momentous and crucial to the destiny of life on this planet than those at Kitty Hawk and Cape Canaveral some 350 million years later.

At the time of this evolutionary breakthrough, Earth had been supporting life in various forms for more than 1,000 million years. Flightless insects had been sharing fern pond and mud flat with scuttling millipedes, spiders and scorpions for perhaps 25 million years. But with the arrival of the flapping wing, insects gained an advantage over all other creatures, one that would lead to a species and population explosion as yet unmatched in size and duration.

Unquestionably, the aerial performance of insects is spectacular by any standards, yet far less is known about insect flight than about the flight of birds or airplanes. Moreover, until recently, even the fundamental laws of aerodynamics appeared to break down when applied to insect flight.

THE EVOLUTION OF WINGS

The evolution of the first wings took place at roughly the same time that the first amphibians abandoned the water for a life on land. Insects and other established earthbound creatures were probably a convenient source of food for the newcomers, so it is possible that amphibians provided insects with the evolutionary challenge which eventually led to the development of wings.

There are no fossil records of flying insects until the late Carboniferous period (about 310 million years ago), and any hints of wing evolution before this time are virtually absent. Thus all theories as to the origin of wings are highly speculative. One reason for this lack of fossil evidence may be the fragility of insects, since most disintegrated before becoming fossilized. A more likely reason is that as soon as wings evolved, insects were provided with such an enormous advantage that in evolutionary terms, there was an explosive radiation of new species and wing designs in a very short time.

What would protopterygotes—the immediate ancestors of winged insects—have looked like? Unlike the evolution of other organs, the wings are unlikely to have evolved as wings from small structures, be-

Dragonflies are among the most primitive of modern insects. Some prehistoric species of the Carboniferous period boasted a wingspan of over 30 inches, but nowadays, few exceed about 4 inches. The southern hawker (Aeshna cyanea), FACING PAGE, *shows the complex veined structure of its wings.*

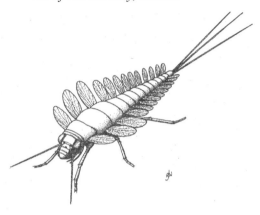

*Figure 35*A. *One theory as to the origin of wings is that they evolved from lateral extensions at the top of the thorax wall, appearing first as shallow flaps which could have served as solar panels. Eventually, these flaps may have become large enough to function as stabilizers when the insect jumped out of harm's way,* RIGHT.

*Figure 35*B.
A primitive insect with winglets.

cause unless they had been large enough to provide some aerodynamic advantage, natural selection could not have improved upon them. The first wings, therefore, must have been preadapted for a completely different purpose.

One theory is that wings evolved from lateral extensions at the top of the thorax wall, appearing first as shallow flaps, which may have served as solar panels by increasing the surface area for absorbing the sun's rays. In the fullness of time, the flaps may have become large enough to function coincidentally as stabilizers when the insect jumped out of harm's way (see Figure 35A).

Supporting evidence comes from several early-insect fossils, which actually exhibit a pair of winglike lobes on the first segment of the insect's thorax in addition to the two pairs of fully developed wings on the other two segments. Perhaps all three pairs started to evolve together, but the front ones never got very far in their development, having been eliminated at an early stage. These early fliers may have been strong jumpers, like the modern grasshopper, or creatures like the cockroach that leapt from the sides of tree trunks to plane away to safety. It is not beyond the imagination to visualize how such prowings developed muscles to control and power movement.

Another theory, made popular by Jarmila Kukalova-Peck of Carleton University, in Ottawa, Ontario, is that insect wings arose from already articulated and movable winglets sprouting from the sides of the thorax and abdomen of aquatic nymphs, like those of mayflies. These winglets would have functioned as gill plates and were used for ventilation and possibly for swimming (see Figure 35B). The aquatically derived winglets must have been small to begin with, but by the time of transition to land, they may well have been large enough to provide an immediate aerodynamic benefit and, as a result, may then have evolved very rapidly.

A virtue of the gill-plate hypothesis is that because the pairs of winglets were situated down the whole length of the body, their total overall area would have been enough to give immediate aerodynamic benefit. Additionally, since the winglets had previously functioned as aquatic ventilating fans or as an aid in locomotion, they would already have some musculature and nerve supply for movement and control. Dr. Robin Wootton of the University of Exeter, an expert on insect wings, and Charles Ellington of Cambridge University have investigated how well such insects might have been able to glide by experimenting with lightweight balsa-wood models whose winglets could be twisted to provide control.

The multiflash photograph of a

model protopterygote made by the author demonstrates how 10 pairs of small winglets can extend a modest jump into a long, controlled glide. Furthermore, Wootton and Ellington found that if the winglets on the abdomen were removed and those on the thoracic part of the model simultaneously enlarged, the models could still glide well—but only if long bristles were attached to the abdomen to maintain stability. Similar long tails are found on modern mayflies and on most of the earliest fossil insects. As the thoracic wings enlarged, so the ad-

vantages of flapping would have increased, and powered flight would have evolved.

Yet another theory, which emerged from the work of James Marden and Melissa Kramer, suggests that small flapping winglets could have been used by semi-aquatic insects to skim along the water surface, as some stoneflies and other insects do today. The progressive winglet enlargement associated with improved skimming efficiency would at last have yielded wings big enough to have lifted the insects into the air.

Figure 36. In some fully flying insects of the late Carboniferous period, like this ancient nett wing, forewings and hind wings overlapped to such an extent that the insects must have looked like flapping biplanes.

Whatever their origin—whether derived from solar panels or gill plates—wings formed the basis of the elegant and efficient mechanisms of today's flying insects. Fossil records indicate that there were a variety of fully flying insects by the late Carboniferous period. They included primitive cockroaches, mayflies and dragonflies, some with wingspans of well over two feet, as well as a wide range of forms in an extinct group known as the Palaeodictyopteroids, which often had archaic net-veined wings. In some of these, the forewings and hind wings overlapped to such an extent that the insects must have looked like flapping biplanes (see Figure 36).

The Permian period heralded the spread and diversification of the reptiles and was a prolific time for insects, with grasshoppers, leafhoppers and several other familiar insects taking to the air. Modern pollinating insects did not evolve until flowering plants started to appear about 130 million years ago. It was then that moths, flies, wasps and, later, butterflies and bees began to flourish and rapidly assume an ever-increasing variety of aerial forms. After millions of years of the search for and discovery of new habitats and food supplies, the enormous number of highly specialized insect species of today came into being, evolving complex life histories and behaviors that can be described only as awesome.

THE INSECT'S SPECTACULAR SUCCESS

Of all the creatures in the animal kingdom, the insect is the most successful. As well as outnumbering by five times the other known animal species put together, insects have invaded and dispersed to every corner of the world, their numbers multiplying as the climate becomes hotter. Although insect success can be attributed to a number of factors, the conquest of the air is the most important. With the exception of birds and bats, the ability to fly sets insects apart from other creatures. If conditions become unfavorable in one place—if food is in short supply, for instance, or enemies threaten—insects simply take to the air and find a more hospitable habitat elsewhere.

At least three other major factors have contributed to the insect's success: its small size, metamorphosis and an external skeleton. First, a crumb too tiny to be noticed by a larger animal can provide a feast for an insect, a drop of dew will quench its thirst, and a fallen leaf will protect it from the midday sun. Many insects are able to occupy minute habitats, such as the inside of a seed or the tissue between the upper and lower surfaces of a leaf, which are obviously unsuitable for larger animals.

Second, the majority of insects benefit from having what is known as an incomplete metamorphosis: Their lives are divided into two active stages, in which they feed on entirely different foods. In this way, an area can support far greater numbers than it could if the insect consumed only one type of food throughout its life. Take the leaf-eating caterpillar and the nectar-sipping butterfly, for example. Growth occurs only in the caterpillar, or larval, stage, during which the creature simply becomes a superefficient organism for gathering food. After the dormant intermediate pupal stage (the chrysalis), the insect is transformed into a butterfly attuned to a life of reproduction. The mature insect, complete with wings and reproductive organs, can now devote the remainder of its life to finding a mate and laying eggs in an area favorable for the growth of its offspring.

The insect's third asset is its hard yet lightweight outer skeleton—the exoskeleton, or cuticle—which contains chitin and serves as a protective armor and an attachment for muscles. The chitin provides fibers in a horny matrix that acts like the resin matrix in the superfiberglass or carbon-fiber man-made composites used in tennis rackets and stealth bombers. As well as providing enormous strength, the cuticle is an efficient two-way waterproofing system: Not only does it keep water out, but more important, it protects the insect from dehydra-

tion. To improve the efficiency of the waterproofing, the outer surface of the cuticle is covered with a thin layer of wax. The problem of drying out is particularly severe in small animals, such as insects, because of their large surface area (see page 30). Consequently, the loss of water by evaporation becomes relatively much higher as the animal becomes smaller.

Together with flight, these three factors combine to give insects a degree of adaptability unheard of in any other form of animal life, allowing them to take advantage of almost any conditions on Earth. They have been found flying at over 15,000 feet, and in one West African species of midge, the larva can survive not only being frozen in liquid air at minus 310 degrees F but also being boiled in water for short periods. In addition, insects have exploited practically all possible food supplies, from green plants and roof timbers to paintbrushes and pools of crude petroleum.

FORM & FUNCTION

Insects have a greater variety of form and function than do any other animals, but their basic structure is uniform. Compared with a bird, an insect is an incongruous creature. Apart from wearing its skeleton on the outside, its body is made up of a series of segments connected to one another by flexible joints. An adult insect is divided into three main sections: the head, the thorax and the abdomen, each with its own functions (see Figure 37). The head carries the mouthparts and principal sensory organs, such as the antennae and eyes, and is mainly responsible for orientation and feeding. The thorax, with its

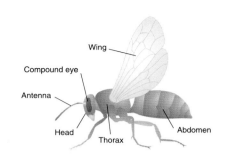

Figure 37. Insect structure.

Compared with that of other flying animals, the external structure of insects appears complex. But unlike birds, bats and aircraft, most insects do not need to be streamlined, since few fly faster than around 20 miles per hour. The various structures protruding from the body of this scorpionfly (Panorpa *sp*) *do not seem to hamper its flight.*

massive musculature, is almost exclusively concerned with locomotion. Three pairs of legs are attached to the thorax and usually one or two pairs of wings. The abdomen contains the digestive, excretory and sexual organs, together with the insect's long, tubular heart and most of the breathing mechanism.

These basic characteristics distinguish the insect from its arthropod relatives, such as spiders, crabs and centipedes, which all boast more than three pairs of legs (spiders have four) and a body divided into either

two main segments or more than three. However, as in other arthropods, the insect's blood system has few vessels; most of the blood is contained in large cavities throughout the body, where it slowly circulates, freely bathing the various organs. The blood has no red corpuscles and plays no part in carrying oxygen around, its main functions being to clear the body of bacteria and particles from cell breakdown and to transport fuel, hormones and nutrients to the tissues.

Since the blood does not absorb

oxygen, an insect does not need lungs and so has evolved a totally different method of respiration from that of mammals and birds. Air seeps into the insect's body through small openings in the cuticle, called spiracles, and is distributed to all parts of the body through a maze of microscopic air tubes known as tracheae, which ramify throughout the tissues. The oxygen is not taken in through the walls of the tracheae but is absorbed in fluid at the ends of their finest branches. It is probable that these minute branches actually penetrate most of the cells of the muscles and other tissues to ensure a sufficient and accessible supply of oxygen.

It was once thought that air could be conveyed through the tracheae solely by the simple process of diffusion, and as diffusion worked only over short distances, insect size was therefore limited. We now know that the majority of insects do not rely entirely on this simple form of respiration but have a ventilating system which forcibly drives air through their bodies, similar to the way in which we breathe. Some of the tracheae open out to form large thin-walled air sacs that collapse and expand as the result of a slow, rhythmic muscular action by the abdominal segments. If you examine a resting hover fly or wasp closely, such movements can be seen clearly.

Furthermore, the main flow of air in certain insects is unidirectional. When the abdomen expands, air enters the body through one set of spiracles, and when the abdomen contracts, it passes out through another, the rate of ventilation increasing as the insect's activity becomes more vigorous. By far the most strenuous exercise is flight, and here the considerable pulsating distortions of the thorax caused by the flight muscles act as a pump. In this way, the thoracic air sacs are made to function like bellows, driving the air in and out of the tracheal system and thereby keeping the rate of oxygen supply to the flight muscles in pace with the insect's flying activity—a simple but effective mechanism.

CONTROLLING TEMPERATURE FOR FLIGHT

Before an insect can take to its wings, its body temperature must be sufficiently high; in the majority of insects, the flight muscles are unable to develop full power until they are at least 75 degrees F. Birds have automatic mechanisms for maintaining body temperature and are therefore not limited by the temperature of their surroundings, but with an insect, the body temperature can, and normally does, rise and fall according to that of the air. Physical activity is related to these conditions, so when an insect is cold, it becomes slow in movement, even torpid.

Thus most insects are active only when ambient conditions are hot enough, often relying on the sun's radiant heat to keep their bodies at a suitably high temperature for flight and mysteriously vanishing from sight if the sun disappears behind a black cloud. When the sun reappears, the smaller insects take to their wings first, since they warm up faster than the larger species because of their higher surface-to-volume ratio. Many insects can, to some extent, actively regulate their temperature by controlling what areas of the body are exposed to the sun's rays. Grasshoppers, for ex-

As well as possessing powerful flight muscles, the hawk moth appears beautifully streamlined, although the advantage of such streamlining is debatable, as this handsome insect is not the fastest of insects on the wing. In fact, it is unlikely that any insect flies fast enough to benefit significantly from streamlining. The thorax of the hawk moth is packed with powerful flight muscles. By shivering its wings before takeoff, the hawk moth can raise the temperature of these muscles to over 85 degrees F. This ravishing Venezuelan cloud forest species (Xyloplanes pluto), RIGHT, *is delicately marked with shades of olive-green and silver.*

Although essentially "cold-blooded," the emperor moth (Saturnia pavonia), FACING PAGE, *raises the temperature of its flight muscles some 45 Fahrenheit degrees above that of its surroundings before taking off. The large, feathery antennae of this male can detect the scent of a female several miles away.*

ample, face the sun during the heat of the day but turn broadside when the sun is low. Butterflies, too, orient themselves so that their wings are angled to the sun to absorb an appropriate amount of heat.

Large-bodied insects, such as silk moths, hawk moths and bumblebees, have found another way of raising the temperature of their bodies that makes them somewhat independent of their surroundings. They use the metabolic heat generated by their flight muscles, which they warm up by shivering their wings before takeoff. In this way, the temperature of the flight motor is raised to 85 to 105 degrees F, which may well be much higher than the temperature of the surrounding air.

This action can often be seen in moths that are disturbed at temperatures too low for flight. The emperor moth, for example, can raise the temperature of its thorax some 45 Fahrenheit degrees above that of

its surroundings. The bumblebee, on the other hand, is capable of flight when conditions are only one to two degrees above freezing, reaching an internal temperature of over 85 degrees F. For such a tiny creature to generate so much heat requires an enormous expenditure of energy. Even more remarkable is the bumblebee's ability to uncouple the flight muscles from its wings, vibrating and warming them up without actually moving the wings at all.

It is essential that insects maintain the temperature of their flight motor, so it is hardly surprising that the thorax of most species is insulated by a dense coating of hairs or scales, as can be seen in butterflies, moths and bees. Even the almost bare shining thorax of the dragonfly is insulated by a layer of air sacs just beneath the outer cuticle. It is also important for insects to take precautions against overheating while flying, as flight muscles that become too hot are unable to function. One

The monarch butterfly (Danaus plexippus) *has to convert nectar into fat before embarking on its long migratory flights.*

of the ways that an insect does this is by passing blood back from the thorax into the abdomen, where it is cooled before being returned once again to the flight muscles.

FUEL FOR FLIGHT

The amount of energy used by flight muscles during periods of activity is extremely high. In fact, the metabolic rate of the fibrillar muscles of flies and bees (see page 53) is about 10 times that of the human heart. The fuel and oxygen consumption of this most active of animal tissues is so great that blowflies may lose up to 35 percent of their body weight in one hour's flying.

Fat serves as fuel in the majority of insects, while carbohydrates are used by others. Both have pros and cons. In terms of weight, fat is far more efficient than carbohydrates, but it suffers in the comparison, because before fat can be used, it must be broken down and transported from storage to muscles. The chief advantage of carbohydrates is that they are water-soluble and are therefore able to circulate in the blood, which bathes the muscles, and so provide instant energy.

The water solubility of carbohydrates, however, brings with it certain disadvantages, because an insect can dissolve only a limited amount of sugar in its blood, further limiting the supply of potential energy that it can carry around. The normal concentration of blood sugar in the honeybee, for example, is 2.6 percent, but once this level falls below 1 percent, the bee is unable to remain airborne and is forced to refuel with nectar.

Thus short-haul insects, such as bees and flies, rely on frequent topping up with sugar, while insects that have to migrate or fly for sustained periods, such as locusts, burn fat. Although butterflies and moths feed on nectar, they have to convert it into fat before they can store it as a source of energy.

THE WINGS

Fully developed insect wings are possessed only by the adults and, having evolved from different structures, are entirely different from those of birds. All birds' wings are substantially similar to one another, but insect wings exhibit an enormous variety of size and design, ranging from the hair-fringed club-like extensions of minute thrips to the transparent membranous wings of dragonflies and the scale-covered wings of exotic butterflies.

Furthermore, unlike the wings of birds, which are modified forelimbs and so have their own muscular structure, insect wings contain no muscle or tendons. They are merely aerodynamic surfaces controlled and powered from the thorax. Another difference is that insect wings are much thinner and flatter than birds' wings and, in fact, do not resemble typical airfoils at all. Nevertheless, once insect wings start flapping and air begins to flow around them, they change shape and become cambered. Dr. Robin Wootton has shown that the nature and extent of this change is not haphazard but is determined by the varying architecture of the

wing, which bends along specific lines or furrows to gain the optimum aerodynamic effect.

Insect wings grow out of the second and third thoracic segments and are made up of two layers of chitin that are sandwiched together and strengthened by a network of tubular veins, many of which contain air-filled tracheal tubes and blood. Most of the wing is dead, but because some of the hairs act as sensory organs, the wing is furnished with nerve fibers. Although many of the veins contain blood, which is particularly important for the wing's development after the insect hatches from its pupa, their chief function is to provide strength.

Little is understood about the ultrastructure of the chitinous membrane, but its strength and lightness are quite extraordinary, even for insect cuticle, which is already a superb natural composite. The wing membrane of some insects is a mere micron or so thick, yet it can withstand the powerful lift forces generated during spectacular aerial

This scanning electron micrograph image of a damselfly wing shows the amazing three-dimensional structure that appears to be two-dimensional.

*The forewings and hind wings
of bees and wasps are coupled
together in flight.*

more advanced insects, such as
bees and flies, is more simplified.
Most insect wings are designed to
twist extensively so that they can
generate lift on the upstroke as well
as the downstroke, allowing slow
flying and hovering to be effected
more easily and efficiently (hovering in most animals, such as birds
and bats, requires very extensive
wing movement and is costly in
terms of energy). Insects such as
flies and dragonflies fall into this
category and possess wings that are
miniature masterpieces of ingenious
design. In direct response to aerodynamic loading, these wings have
evolved a battery of elegant mechanisms that automatically maintain
an ideal camber and angle of attack.

Ake Norberg of the University of
Gothenburg in Sweden has demonstrated the importance of preventing wings that readily twist from
pitching up into the wind like a flag.
He showed that the pterostigma, a
fluid-filled cavity near the tip of the
leading edge of many insect wings,
acts as a counterweight to oppose
this pitching motion. Insects such
as flies also possess leading-edge
veins and several other veins leading toward the wingtip which are
designed to twist torsionally in such
a way that the wing adopts a pro-
peller-bladelike shape and produces
a cambered profile. Other similar
mechanisms operate whether aerodynamic forces act in the wing ven-
trally, as on the downstroke, or dor-
sally, when the wing twists around
for the upstroke. Such devices are
ideal for insects that invert their
wings and reverse camber.

Originally, the earliest flying insects had two pairs of wings, each
pair functioning independently in
flight, a system that has survived up

maneuvers. Whether it be digital
coding, zip fasteners or lattice gird-
ers and space frames, nature always
seems to get there first. Look at the
corrugated structure of the damselfly wing on page 47. Its enormous
strength can be gauged with a quick
glance by a scanning electron micrograph. The thin membrane between the cross-veins is sheared,
contributing to rigidity by bracing
the structure diagonally.

In flight, maximum stress occurs
around the leading edge, so the
veins here are thicker and closer together. The wings of more primitive insects are further reinforced
by fine veins crossing the longitudinal ones, but as the evolutionary
tendency has been to reduce the
lateral veining and to strengthen
the longitudinal, the veining of the

A cockchafer *(Melolontha melolontha) has just taken off. As well as protecting the wings at rest, the elytra of beetles probably function as fixed airfoils in flight.*

Figure 39. The transparent wings of bees and wasps are coupled by an elaborate system of hooks on the leading edge of the hind wing that engage on the forewing's trailing edge.

to the present day in insects such as the dragonfly. But generally, greater efficiency is achieved when the two pairs of wings beat together. Most insect orders have shown an evolutionary trend toward a single pair for all practical purposes.

Four-winged insects, such as butterflies, moths, bees and wasps, approach a "two-winged condition" by coupling their forewings and hind wings together. In butterflies, the leading edge of the hind wing has a lobe that locks under the overlapping forewing. Moths link their wings by means of a stiff bristle or group of bristles (the frenulum) at the base of the hind wing

The drumsticklike halteres of the crane fly and other flies help to orient, or "balance," the insect in flight, perhaps acting a bit like the turn-and-bank indicator of an aircraft.

The flea uses its "old" flight muscles to power its prodigious leaps.

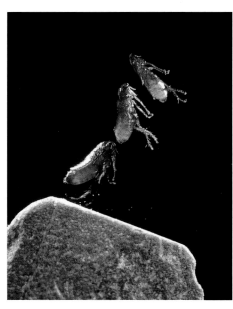

that hooks onto a catch (the retinaculum) on the underside of the forewing. The transparent wings of bees and wasps (Hymenoptera) are coupled by an elaborate system of hooks on the leading edge of the hind wing that engages on the forewing's trailing edge (see Figure 39).

The beetle (Coleoptera) has followed this trend, but in a different way, since only its membranous hind wings take an active part in flight, the forewings having become modified as protective covers called elytra. During flight, the elytra are held at about a 45-degree angle above the body, functioning in a limited capacity as fixed airfoils to increase lift (see photo on page 49).

Associated with wing coupling and the two-winged condition is a general tendency for the hind wings to become smaller, which in true flies (Diptera) has resulted in the hind wings being discarded completely. All that remains is a pair of drumsticklike knobbed stalks known as halteres, which help to orient the insect in flight.

Where insects have found wings

more of a hindrance than an aid to their way of life, evolution has seen fit to go backward and discard the wings altogether. The loss of wings has occurred in species from most orders, including moths, beetles and flies. While the wings of ants would be an encumbrance only in their cramped underground quarters, the workers are wingless from the outset and the males and queens rid themselves of their wings immediately after mating. Certain parasitic flies have also lost their wings, as they are hardly needed for rummaging around the fur and feathers of their hosts. The females of some moth species are wingless, since they spend their lives laying eggs close to where they themselves were hatched. Fleas, too, abandoned flight millions of years ago, their flight muscles becoming adapted for powering their hind legs to effect their prodigious leaps.

THE FLIGHT MOTOR

The wings are attached to the insect's thorax, which is packed with a complex of flight muscles together with various other structures for coupling and operating the wings and, to a lesser extent, the legs. It is from here that all the power and control emanate, enabling the insect to carry out almost any aerial maneuver. How the power is applied from the thorax to the wing varies from one insect order to another, but basically, the wings are hinged so that they can move freely in almost any direction, the coupling system functioning like a series of ball-and-socket joints. The thorax can be considered simply an elastic box with a lid that is held in place by a membrane coupling it to the top of the sides. The wings pivot on the underside of their bases and are hinged onto the thorax lid on the upper side.

In the more primitive insects, such as dragonflies and grasshoppers, the downward movement of each wing is produced by a powerful flight muscle that runs from the thorax floor directly to the base of the wing beyond the pivot point. When this muscle contracts, the wing moves downward. Upward wing movement is powered by a flight muscle that runs from the floor of the thorax to the inboard side of the wing's pivoting point. Contraction of this muscle pulls the root of the wing down into the thorax, effecting an upward movement (see Figure 40).

With such an arrangement, each of the four wings has its own independent power supply provided by the two direct flight muscles. In this way, the two wings on each side need not necessarily beat in unison, as can be seen in the photograph of the damselfly (see overleaf).

The front and rear pairs of wings in damselflies can be, and usually are, out of phase by as much as half a cycle. The system works well for insects that do not have a wing-beat frequency much higher than about 25 beats per second, as the nervous impulses which initiate the up-and-down movements of the wings are also working at the same slow rate.

Many insects, such as bees and flies, have a wing-beat frequency of 10 times this figure, and as their nervous systems are unable to keep pace with such high speeds, they have solved the problem in another way. To begin with, the thoracic lid is less flexible, so when the flight muscles contract, the whole thoracic box is compelled to change

shape as a unit. These insects have two pairs of muscles that, rather than being attached near the wing's base, are connected to the walls of the thorax (see Figure 41).

One pair, the indirect vertical muscles, runs from the roof of the thorax to the floor, while the other pair, the indirect horizontal muscles, runs horizontally, connecting the rear to the front. The wings are coupled to the sides in such a way that as the thorax changes shape, the wings move. Thus contraction of the vertical muscles pulls the lid of the box down and forces the sides out like a squashed rubber ball, resulting in the wing's movement. Due to the system of leverage, just a small movement at the wing's root causes the wing to move over a relatively large distance.

When the longitudinal muscles contract, the box squashes in the opposite direction, arching the roof of the thorax upward, which causes the wings to move on their downward stroke. It is interesting to note that in insects, the muscles which create the upstrokes and the downstrokes are the same size, and both deliver power. In birds, the upstroke is relatively weaker, the muscles that produce it being smaller.

Another factor effecting high-frequency wing movement is the mechanism used for wing articulation. Its development reaches a peak in flies and in bees and wasps, the two insect orders with higher wing-beat frequencies than all others. The wing couplings were once thought to act like a light switch, with two stable positions, either up or down.

This "click mechanism," as it was called, was credited with accelerating the wings up and down

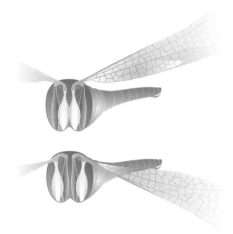

Figure 40. The direct muscle-powered flight motor of the dragonfly. In the upper diagram, a pair of direct flight muscles contracts, resulting in an upward movement of the wings. In the lower diagram, the other pair of muscles contracts, resulting in the wings' downward motion.

Figure 41. The indirect muscle-powered flight motor of a fly. Vertical muscles are shown contracted in the upper diagram while the horizontal muscles are contracted in the lower diagram.

through the stable position with a clicking action. However, recent research by Roland Ennos at the University of Exeter and others suggests that the click mechanism does not really exist but was, instead, an effect of the anesthesia used in earlier experiments, which changed the physical characteristics of the tissues involved.

Although the wing action is brought about by the thorax, it is the combined effect of the thorax's elasticity, the "double-jointed" wing coupling and the movement of the lateral scutum (top side of thorax) that helps to "flick" the wings up and down (see Figure 42). Thus an upward and outward movement of the lateral scutum initiates the downstroke, while a downward and inward movement returns the wings to their upper position.

The efficiency of an oscillating system, such as an insect's flight motor, is substantially improved by the great elasticity of the thorax, allowing it to resume its shape spontaneously after each contraction. Once set in motion, the flight motor needs sufficient energy only to replace that which is lost due to wind resistance or is absorbed by friction. The insect's thorax contains many elastic energy-storing tissues, the most remarkable of which is the wing's minute suspension system. In many insects, this is made of resilin, the most nearly perfect elastic material known and one that so far has defied man's attempts at synthesizing.

Together with the development of the method of articulation, insects with high-frequency wing movement have evolved an extraordinary muscle tissue known as fibrillar muscle. Probably the most

A damselfly (Hetaerina cruenta) *from Venezuela,* FACING PAGE, *clearly shows the independent movement of the two pairs of wings that are out of phase by about half a cycle.*

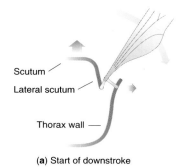

Scutum

Lateral scutum

Thorax wall

(a) Start of downstroke

(b) Start of upstroke

Figure 42. A transverse section of a fly's thorax shows how the indirect flight muscles acting on the thorax walls cause movement and flexion of the lateral scutum (top side of thorax) and a corresponding articulation of the double-jointed wing-coupling mechanism, which in turn is converted into wing motion. In (a), the scutum is moving up, which pulls the wing down; in (b), downward movement of the scutum pushes the wing up. Note that the joint fulcrums do not overcenter at any time.

A multiflash photograph of a yellow dung fly (Scatophaga stercoraria) *as it leaves a rhododendron seedhead. The wings of a fly are attached to the thorax, which acts like a pulsating box. When the fly takes off, its flight motor is stimulated into activity automatically by special "starter" muscles connected to the center pair of legs.*

active tissue ever to have developed in a living organism, fibrillar muscle has the peculiar quality of automatically contracting rapidly after being stretched. Once impulses from the central nervous system initiate the process, the power supply for the wings runs by itself until the impulses stop. The muscles can contract and relax far more rapidly than the nerves can fire them; hence the wing-beat frequency is determined by the resonant characteristics of the wings and thorax rather than the lower rate of nerve impulses. Through a combination of such refined mechanisms, many flies, bees and wasps are able to beat their wings at around 250 cycles per second (one cycle is a complete up-and-down movement), while some midges can flap their way through the air at a staggering 1,000 strokes per second.

As well as beating up and down,

the wing must be extensively moved in the other two planes during flight. For instance, the inclination of the wing must be adjusted so that it can be twisted along its axis. This is accomplished by tiny accessory muscles attached to the roots of the wing, while other muscles effect the wing's forward and backward motion. Another set of small muscles helps to control the elasticity of the thorax by bracing the walls against the strain of the main flight muscles. These active adjustments, along with a complex of inertial and aerodynamic forces, combine to twist and camber the wing to achieve the lift and control necessary for the remarkable aerial performances we take for granted.

Being the anthropocentric creatures we are, it is easy for us to assume that small animals, like insects, are simpler than large ones, because they have less inside them.

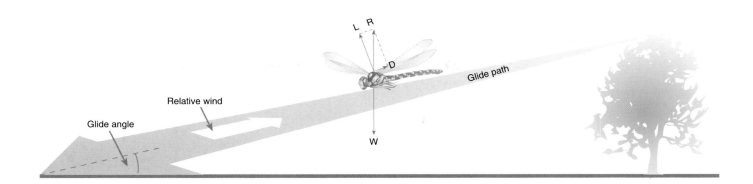

Figure 43. The forces that act on a gliding insect: its weight (W), its lift (L), the drag (D), the resultant upward force (R) together with the glide path and the relative wind.

This is a serious misconception. Apart from the development of the brain, the fly is almost as complicated and miraculous as man, possessing everything necessary for its special way of life, from an intricate tracheal system to a thorax packed with sophisticated tissues and mechanisms for operating the wings. Yet, without a second thought, we can destroy it at a single stroke!

GLIDING FLIGHT

During steady forward flight, the four forces acting on an insect (weight, lift, thrust and drag) are in equilibrium (see Figure 9 on page 20), but when the insect is gliding, no thrust is generated—its wings have stopped beating. Now, in order for the insect to remain airborne and under control, it must derive power from the force of gravity. It gains the airspeed necessary to maintain lift by inclining its head downward relative to the air (see Figure 43). In so doing, the insect moves forward but steadily loses height; the more efficient the gliding, the shallower the glide path (see "glide ratio" on page 94).

It is striking that only a few large insects have exploited the simple technique of gliding. After a period of conventional flapping flight, some dragonflies will sail on motionless outstretched wings for several seconds without losing too much height before resuming normal flight. Some large butterflies, such as the monarch, habitually glide for short distances, and locusts have been known to ascend in thermals to over 3,000 feet with barely a flap of their wings.

These insects glide by relaxing their flight muscles so that the wings lock into an appropriate shallow V attitude for maximum lateral stability, allowing them to sail through the air without any muscular effort. Most insects, though, make inept gliders because of high friction drag and a correspondingly poor lift-to-drag ratio, which is in contrast to many bird species whose survival depends on their gliding ability.

FLAPPING FLIGHT

Flapping flight is obviously far more complicated than gliding, and many of its subtleties are poorly understood. It was once thought that insects and birds "swam" through the sky by rowing their wings against the air, but we know now that this is not so. The wings of insects and birds act like airfoils, each section meeting the air at an angle of attack and generating varying degrees of lift and drag. Such flight depends on a complex combination

Some species of butterflies are among the few insects capable of gliding. Members of the Nymphalidae family, such as the white admiral (Limenitis camilla), RIGHT, *can often be seen gliding down from the trees, albeit at quite steep angles.*

A lacewing (Chrysopa *sp*), FACING PAGE, *caught a split second after take-off as it wings its way upward like a miniature helicopter.*

of wing-flapping and wing-twisting and varies according to the species and the type of flight.

Unlike airplane wings, which provide only lift, the wings of flying animals must generate thrust as well, functioning more like propeller blades. A propeller drives a blast of air behind it, creating a high-pressure area behind and a low-pressure area in front. It is like a wing to the extent that it generates both lift and drag, but the difference lies in the fact that the lift is directed forward. Nevertheless, the lift is still that part of the force (on the propeller) which is at right angles to the blades' motion.

Whereas the plane of movement of an airplane's propeller is at right angles to the machine's horizontal axis, an insect's wing is not confined to any plane. It starts its action from a point above and behind the thorax, and during the downstroke, it moves forward as well as downward. In reality, it follows an

ellipse or a slender figure eight, depending on the nature of the flight (see Figure 44).

The combined effect of these actions is to fan a current of air downward as well as backward, providing both lift and thrust, so the insect not only moves forward but is simultaneously supported against the force of gravity. When the insect wishes to climb, all it need do is adjust the angle of its wing beat so that the current of air is directed more downward and less backward.

The flight of an insect is more akin to that of a helicopter than to that of a fixed-wing airplane, in that both propulsion and lift are provided by moving wings. A helicopter in level flight moves with its tail end tilted upward so that the rotor blades are inclined to the direction of motion in a manner similar to that of an insect's wings in level flight (see Figure 45). Both rotor blades and flapping wings move forward and downward, then up-

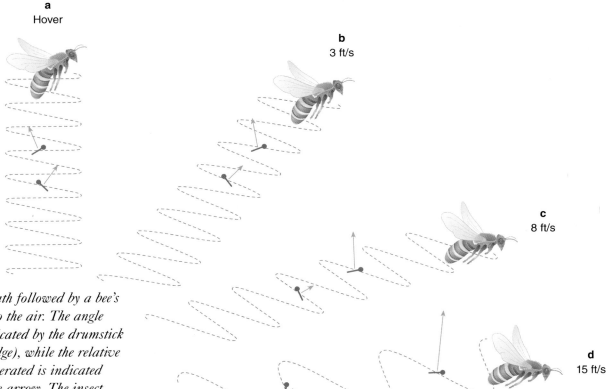

a
Hover

b
3 ft/s

c
8 ft/s

d
15 ft/s

Figure 44. The path followed by a bee's wingtip relative to the air. The angle of the wing is indicated by the drumstick (blob is leading edge), while the relative amount of lift generated is indicated by the length of the arrow. The insect is shown flying at four speeds, from 0 feet per second (hovering) to 15 feet per second.

Figure 45. The inclination of insect and helicopter in forward flight and the resulting downwash.

ward and backward, and both fan a stream of air obliquely downward, analogous to the downwash created by the fixed wings of an airplane. During hovering flight, the helicopter rotors assume a horizontal plane so that air is drawn from above and flung away downward. Likewise, most insects use a near-horizontal stroke when they hover.

The wings of birds and insects achieve a similar effect to that of the propeller by adjusting their angle of attack with each phase of the wing beat. Let us follow the cycle of wing movement of a fly (see Figure 46). From the fully raised position, the wings move forward and downward, accelerating as they do so and attaining a maximum speed in the midposition. At the start of the movement, the angle of attack is extremely high—about 90 degrees—but this is rapidly reduced as the

wings sweep downward, reaching their minimum in the center (see Figure 46c). Once past the central position (quarter-cycle), the wings slow down and begin to rotate—root first—in the opposite direction, with the angle of attack increasing again (see Figures 46d and 46e). Thus the maximum angle of attack occurs when the wing is moving at its minimum speed.

At the bottom of the stroke, when the speed is at its lowest but the wing is still rotating, the wing is suddenly moved obliquely backward and upward (see Figure 46f). However, it continues to rotate around its longitudinal axis until a stage is reached when the upper surface of the outer two-thirds of the wing may have twisted so much that it faces downward (see Figure 46g), not easily shown in the diagram but well demonstrated in the photographs of the mud dauber wasp, below, and the parasitic fly (see page 62). The degree of twisting depends on the type of flight. The upstroke continues at an oblique angle, so the wing traces out a path behind that of the downstroke.

As the wing reaches the end of its upstroke, the direction of rotation changes once again, increasing the angle of attack. The complex pattern of movement ensures that the wing attacks the relative wind—remember that the whole insect is traveling through the air—in such a way as to derive suitable proportions of lift and thrust at all stages.

The aerial performance of insects varies enormously in terms of wing beat, speed, endurance and capability and is largely geared to the requirements of habit and habitat. Although wing-beat frequency can be measured relatively easily with a stroboscope, it is very difficult to measure the airspeeds of insects accurately, because they rarely fly in a

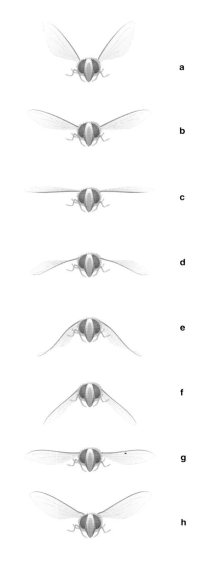

Figure 46. The movement of a fly's wings during the wing-beat cycle.

a

b

c

d

e

f

g

h

The wings of the mud dauber wasp (Sceliphron caementarium), TOP LEFT, *have just begun their downstroke, whereas in the bottom image, the wings are midway through their upstroke.*

In contrast to the lacewing, a bumble-bee, RIGHT, has a high wing-loading and must beat its wings rapidly to remain airborne.

Figure 47.
This chart shows approximate air-speeds and wing-beat frequencies.

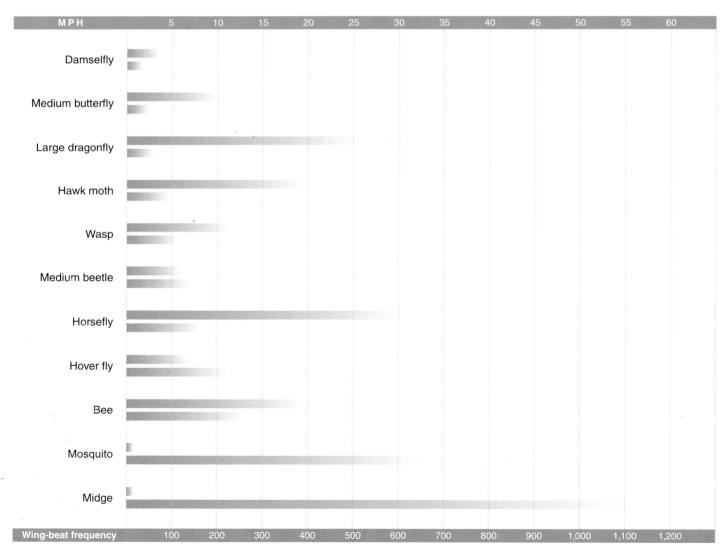

MPH	5	10	15	20	25	30	35	40	45	50	55	60

Damselfly
Medium butterfly
Large dragonfly
Hawk moth
Wasp
Medium beetle
Horsefly
Hover fly
Bee
Mosquito
Midge

Wing-beat frequency	100	200	300	400	500	600	700	800	900	1,000	1,100	1,200

straight line for any length of time. It was once thought that the fastest insect was the hawk moth, as it possesses powerful flight muscles and a beautifully streamlined body, but there is some doubt about this now. Unfortunately, the hawk moth's nocturnal habits compound the difficulty of measuring its speed. Over short distances, bees and horseflies can reputedly keep up with vehicles moving at 30 miles per hour, while some dragonflies are said to be able to do 40 miles per hour for brief stints when pursuing prey. The chart on the facing page shows both wing-beat frequency and approximate airspeeds for various types of insects.

CONTROL & ORIENTATION

Perhaps the most impressive feature of insects is their complete mastery of movement in the air. They seem capable of virtually any flight maneuver, from loops and rolls to flying upside down, sideways and backward to ascending vertically and hovering—including changing from one maneuver to another in a fraction of a second. Watch a dragonfly patrolling a woodland glade as it flies,

Thousands of empid flies dance in aerial courtship over a pond, many in tandem.

An insect's wing movements can be very extensive, particularly when it is taking off or executing erratic turns. In this parasitic fly's (Echinomya fera) efforts to vacate a flower, TOP RIGHT, its wings have rotated 180 degrees, so the trailing edges are now leading.

A South American ithomid butterfly, CENTER RIGHT, displays bizarre wing-twisting as it flies among tropical ferns.

Shortly after a rapid takeoff, this peacock butterfly (Vanessa io), BOTTOM RIGHT, is in inverted flight.

The wings of a broad-bodied chaser (Libellula depressa) *undergo pronounced twisting as it takes off sideways (and toward the camera) from a reed. When the wing movements are examined closely, it can be seen that they are not identical on each side; there is slightly more twisting on the left pair of wings.*

glides, hovers and changes direction almost instantaneously, sweeping in to seize a winged morsel. Study a swarm of gnats as they hang in the air by the hundreds in an ever-changing pattern of evasive action, never colliding (see page 61). Or observe a housefly taking off like a rocket and slipping through the air to land upside down on the ceiling (see page 67). All demonstrate sensational flying techniques that require miraculous control.

The aerodynamic forces involved for altering course are the same for insects as they are for birds, but the techniques used are somewhat different. A bird changes direction in the air by making adjustments to its tail or wings, the degree of movement needed to make the turn depending on its airspeed. A high-speed aircraft requires only minute movements to its control surfaces to effect a substantial change in course, but an insect, because of its lower airspeed, must adopt far more extreme asymmetrical wing positions or movements.

Yet the precise nature of these movements is still obscure, because insects are so difficult to photograph or film in free flight, especially when they are executing erratic turns. At one time, it was even suggested that an insect flaps the wings on each side of its body at different speeds, but we now know this is impossible, especially in those insects with indirect flight muscles, as they cannot power the wings independently.

With insects such as the dragonfly, the wings can be controlled more or less independently by the direct flight muscles, enabling the insect to change course by varying the amplitude of the wing beat on each side of its body. By increasing the amplitude and hence the power of, say, the left-hand side, the insect turns to the right. But how an insect with indirect flight muscles accomplishes a differential amplitude is by no means certain. For effecting small directional changes, accessory direct flight muscles are used for independent rotational adjustment of the wings at their root.

So how does a fly make those spectacular split-second turns? Research in Germany with tethered

The desert locust possesses beds of hairs on its head that are sensitive to air currents and play a part in flight orientation. Locusts and other grasshoppers do not display anything like the aerial virtuosity of bees and flies.

bluebottle flies suggests that the fly may be able to disconnect its wings from the flight motor for split seconds while in flight, allowing one wing to be drawn back into the rest position while the other beats away normally. Although this explanation does sound a bit far-fetched, the magnitude of one-sided aerodynamic forces produced by these means would certainly help to account for the fly's extraordinary maneuverability.

The speed at which flies and other insects are capable of reacting raises an interesting point regarding their timescale. Of all creatures, the fly must be one of the most alert, as it is able to distinguish accurately between events separated by one three-hundredth of a second or less. This vital split second can mean life or death to a fly as it dodges the hand that tries to swat it. Humans are unable to distinguish precisely between events separated by anything less than one-twentieth of a second—over 10 times slower. Our visual system, for instance, allows us to enjoy a film without being aware of the dark intervals between each successive frame, but to a fly, the film would appear like a slide show, with long, dark pauses between each transparency.

Aircraft are equipped with a maze of electronic and mechanical instruments to navigate and regulate their flight. Insects, too, employ a range of detecting and stabilizing mechanisms for orientation and control, about which little is known. Pitch, yaw, roll and airspeed are monitored by sensory apparatus, such as the eyes, antennae and specialized beds of hairs sensitive to air currents.

As far as we know, the fly possesses the most refined flight-orientation equipment in its halteres (see page 50), which act like oscillating gyroscopes, functioning in a similar way, perhaps, to an aircraft's turn and slip indicators. During flight, these oscillate out of phase with but at the same frequency as the wings, while their knobbed ends give them an inertia that tends to keep them vibrating in the same plane. When the fly changes attitude or course, the stalks twist slightly, which is detected by sensory organs at the base. In this way, the fly is made aware of its new flight condition. The dragonfly has developed a flight-orientation system by making use of its heavy freely moving head to detect differences in the position between it and the rest of its body.

TAKEOFF & LANDING

Many insects, particularly those with direct flight muscles, automatically start flying as soon as their feet lose contact with the ground; this is called tarsal reflex. Thus when the insect springs into the air, its wings are stimulated into action. Flies and possibly other insects with fibrillar muscles have special "starter" muscles that run from the legs to the roof of the thorax. These muscles automatically stimulate the flight motor into activity as the legs kick downward, giving the thorax a quick tug. Some indication of the speed and efficiency of this action can be seen in many of the photographs that appear here. However, not all insects are capable of rapid takeoff. Before the beetle can become airborne, for instance, it must first raise its wing cases (elytra), then unravel the wings that are folded underneath—this can take several seconds in the larger bee-

Various stages of the takeoff of a common housefly (Musca domestica).

A cockchafer unraveling its wings prior to takeoff.

gear is underscored by estimates that some beetles are subjected to a force about 40 times the force of gravity when they strike an unyielding surface, like the bole of a tree—such a force would cause the utter disintegration of any aircraft.

A topic of long-standing debate among scientists and laymen alike was the technique used by the fly for landing upside down on a ceiling. Did it perform a roll or a half-loop before touchdown? As it turns out, the truth is both elegant and simple. It does neither, as the multi-flash records for the first time in the photograph on the facing page. It flies toward the ceiling at an angle of about 45 degrees and stretches out all its legs. Using its front legs to touch down, the fly then deftly cartwheels over onto its other four feet to complete the landing.

STALLING

Among the many vexing questions about insect flight is the animal's immunity to stalling. We know that the critical angle for aircraft wings is about 15 degrees, but insects do not stall until they reach much higher angles of attack—up to 60 degrees in the fruit fly. Again, unlike aircraft, which usually stall dramatically by suddenly falling out of the sky, insects lose lift gradually. This gives them a distinct advantage over aircraft, making them less susceptible to stalling in sudden gusts of wind and generally allowing them far more freedom when making abrupt maneuvers.

The way in which insects do this is far from clear, although unsteady airflow effects must be at the root of the answer, as we will see. It is not easy to compare the flight of large bodies with that of small

tles. More will be said about takeoff when we look at the microaerodynamics of insect flight.

Before landing, most insects extend their six legs as soon as they are within a few body lengths of the landing surface. The legs function as efficient shock absorbers since, unlike aircraft and some birds, insects never run forward after touchdown. They arrive at their landing spot from almost any angle without slowing down at all, alighting with a jolt. The efficiency of their landing

ones, because flying characteristics change with differences in scale and surface-to-weight ratio. But as the scale effect is tied up with Reynolds numbers and other mathematical relationships, it need not concern us here. (A brief explanation of Reynolds numbers appears on page 21.)

One critically important factor, and one that certainly affects stalling characteristics, is the behavior of the boundary layer—the thin layer of air in contact with the wing (see page 28). With insects, every vein, hair, scale and corrugation plays a part in the wing's flying efficiency. It has been shown that moths deprived of wing scales generate far less lift than those with scales, but other than that, we know very little about the boundary layer's effects on the microaerodynamics of insect flight.

As mentioned in the previous chapter, insect flight cannot be adequately explained in simple terms, as it is governed by such a vast number of interrelated aerodynamic and biological factors. Furthermore, it is virtually impossible to follow, photograph or measure the multiplicity of fluctuating parameters, such as lift, drag and thrust, while insects are flapping through the air in free flight. But much has been learned recently using large-scale insect models and tethering insects to a sensitive balance in a wind tunnel, enabling filming and aerodynamic measurements to be done more easily. The main area where such research has borne fruit is associated with the very heart of aerodynamic theory: circulation.

NONSTEADY AIRFLOW & THE CLAP-FLING

Conventional aerodynamics is based on steady-state airflow, when a fixed wing or propeller moves through the air at a constant speed, but it cannot explain how insects fly. Indeed, calculations based on conventional aerodynamic theory show that insects should not be able to fly at

Wing scales on a moth's wing, ABOVE. *By affecting the boundary layer characteristics, probably every hair and scale on an insect's wings has its part to play in flight.*

The housefly, ABOVE LEFT, *does not land on the ceiling by performing a barrel roll or a half-loop but by flying up at an angle of about 45 degrees, with its front feet extended. As soon as contact is made, the fly cartwheels over onto its four other feet.*

A lacewing with its wings in the "clapped" position.

ously into the air. An insect just flies—no fuss. It generates lift the moment its wings start moving.

The explanation for the insect's apparent flouting of the laws of aerodynamics brings us to one of the most fascinating aerodynamic discoveries of recent times. It arose from work originally carried out by the late Torkel Weis-Fogh, a zoologist at Cambridge University. After making a number of calculations, Weis-Fogh came to the disconcerting conclusion that certain insects—particularly those with low Reynolds numbers and low wing-loading—are unable to generate sufficient lift to remain airborne, at least according to the accepted principles of aerodynamics.

But insects *do* fly and have been doing so for some 350 million years; therefore, they must derive extra lift by some unknown means. The paradox was taken up by Weis-Fogh, who conducted a series of painstaking observations in the laboratory using high-speed cinephotography of hovering insects. Among the insects studied was *Encarsia formosa*, a minute parasitic wasp often used in the biological control of aphids in greenhouses. As the insect has a wing-beat frequency of 400 cycles per second, the only way to analyze its movements was to film it with a rotating prism camera at 8,000 frames per second.

The first clue to the solution of the mystery was found in a frame-by-frame analysis of the film. It showed that the wasp hovered in the normal manner, with its body vertical and the wings sweeping more or less horizontally. However, at the end of the upstroke, the two pairs of coupled wings "clapped"

all. Somehow, insects produce far more lift than would be expected through normal airfoil action.

What we first notice when watching insects (and, to a lesser extent, birds) take off is that they become airborne without delay, unlike conventional aircraft, which need hundreds of yards of runway to get off the ground, or helicopters, which have to spin their rotor blades for some time before lifting ponder-

over the insect's back (see Figure 48a). Then, after a short pause of one two-thousandth of a second, the wings were suddenly flung open with their hind margins still touching each other (see Figure 48b). Following the "fling," the hind margins separated and the wings moved horizontally through the air, as in conventional hovering flight.

The "clap-fling action," as Weis-Fogh labeled it, occurred in the tiny wasp's every wing-beat cycle, in both hovering and normal flight. Furthermore, the oscillations of the insect's vertical body showed that the lift had equaled the weight of the insect shortly after the clap position, thereby suggesting that air circulation around the wings had been built up a long time before the wings had reached maximum velocity. This is far from what would be expected of ordinary airfoil action.

Rapid takeoff requires extensive wing movement, so in order to establish immediate circulation for lift, an iron prominent (Notodonta dromedarius) *uses the "clap-fling" at the end of both the downstroke and the upstroke (shown here in multiflash).*

| a | b | c | d |

Figure 48. "Clap-fling action."

Figure 49. "Clap-fling-ring action" of the cabbage white butterfly.

After further calculations and inspired thought, Weis-Fogh found a rational aerodynamic explanation. During the clap, the air around the insect is nearly motionless, but as the wings are flung open, air rushes in to fill the growing wedge-shaped space created between the upper surfaces. Thus as soon as the hind margins separate to begin the downstroke, the wings carry a vortex of air formed during the fling. In this way, circulation and lift are established, and the wings immediately begin their downward sweep. An aircraft wing, a propeller or a rotor blade must move several chord lengths before any lift whatsoever is generated.

Since the discovery of the clap-fling action in this parasitic wasp, a similar action has been observed in many other insects, including fruit flies, moths, butterflies and lacewings (see pages 68-69). In lacewings, moths and butterflies, the wings are sometimes clapped at the end of the downstroke as well as the upstroke. Even more intriguing is the fact that the lacewing's two pairs of wings, which are controlled by direct flight muscles, do not necessarily beat together—they sometimes clap 90 degrees (a quarter-cycle) out of phase with each other.

In the case of the cabbage white butterfly (*Pieris brassicae*), the clap-fling works in an entirely different way, as has been demonstrated by Charles Ellington, Weis-Fogh's successor at Cambridge. This butterfly (and probably many other species) has a unique form of flight: It appears to blow a series of "smoke rings" that are left trailing in its wake. Ellington calls this the "clap-fling-ring action" (see Figure 49).

During takeoff, hovering and slow flight, the downstroke always begins with the wings clapped. Rather than being flung open about the hind margins, however, the wings are flung open about the body, which remains nearly horizontal as the wings move vertically downward (Figure 50A).

Instead of circulating across the wing from leading to trailing edges, the bound vortices created by the fling move spanwise and are shed at the wingtips. So, unlike normal airfoil action or the clap-fling, lift is not derived through a Magnus effect but is generated in the following manner.

At the end of the downstroke, the vortices are shed to form one large vortex ring. The fling generates a vortex of air directly beneath the insect, while the force needed to create the ring sustains the insect's weight. Thus the butterfly does not

An unusual view of a cabbage white butterfly (Pieris brassicae) *as it flies away from the camera shows the wings clapped at the end of the downstroke.*

Figure 50A. A butterfly leaves invisible trails of rings of circulating air in some way analogous to the wingtip vortices of an aircraft.

Unlike most other hovering animals, a hover fly beats its wings up and down through a small stroke angle while keeping its body more or less horizontal, as in normal forward flight. Only recently has a satisfactory explanation been offered as to how the hover fly gains sufficient lift while hovering.

fly in a conventional aerodynamic way at all but obtains an upward reaction from the vortex rings that are flung downward. The effect is analogous to the downwash of normal airfoil action, but instead of being created as a steady flow, it is produced in an interrupted sequence. It is interesting to note that the shape of the butterfly's wings, with their low-aspect ratios and consequent high level of induced drag, is unsuitable for the efficient generation of lift by normal airfoil action but is perfect for this function.

Moreover, the erratic up-and-down motion so characteristic of the flight path of many butterflies and moths—especially those with low-aspect ratios and low wing-loading—can now be explained, as the timing of these jerky movements corresponds to the shedding of the vortex rings. How fascinating it would be to see in slow motion the actual air currents and vortices generated by these insects.

THE PARTY-STREAMER EFFECT

Although the discovery of the clap-fling and clap-fling-ring actions sheds light on certain aspects of insect flight, it does little to explain how the majority of insects fly. Most insects, such as dragonflies and large flies, beat their wings obliquely up and down through a small stroke angle, while their body axis remains horizontal. There is no question of any fling action, yet calculations reveal that normal steady-flow airfoil action is quite incapable of providing the insects with sufficient lift to keep them airborne. There must be some other means

by which these insects establish air circulation around their wings, but the details remained a mystery until the late 1990s, when Ellington and his team at Cambridge discovered a new phenomenon of nonsteady airflow that had eluded our understanding of animal flight for so long.

A clue to the newly discovered mechanism at work is provided by watching a simple paper airplane at the point of landing: The plane appears to gain a momentary boost in lift shortly before touching down. This action has nothing to do with ground effect, explained on page 98, but is the result of a delayed stall and occurs when sharp-edged wings cut through the air at high angles of attack. It is caused by air breaking away from the wing's leading edge and somersaulting over it-

self to form a vortex. With its fast-moving swirling air, this leading-edge vortex (LEV) creates the low pressure on the wing's upper surface that generates the final boost in lift to the paper airplane shortly before touchdown.

Unfortunately, perhaps, the delayed stall can boost lift for only a brief moment in conventional aircraft, as the rotating air cannot swirl for long before the LEV rapidly becomes unstable and tumbles away from the wing's surface. Ellington realized that such unsteady aerodynamic effects might play a crucial role in insect flight. Later experiments with a flapping-wing insect model using colored helium-filled bubbles revealed that insect wings could generate LEVs, but most astonishing was that during the downstroke, the vortex is pulled

*Figure 50*B. *A hawk moth shows leading-edge vortices being drawn out to the wingtip.*

It is very likely that leading-edge vortices are responsible for much of the lift generated not only by insects but also by other flying animals. The wings of a silver-washed fritillary (a butterfly with a fairly shallow wing beat), FACING PAGE, *are halfway through their down-stroke, with the leading-edge vortices still attached to the leading edge—the "party streamers" have to be imagined!*

out along the leading edge like a party streamer (see Figure 50B).

By so doing, the vortex is sucked out into the tip vortex instead of growing so large that it breaks away from the wing. It remains stuck to the surface until the wing is well beyond the halfway point of the down-stroke. Later calculations revealed that this additional lift-producing mechanism generates more than sufficient extra lift to sustain the insect in flight. Interestingly, it also permits a wing to travel at a high angle of attack for a brief period before stalling. The phenomenon certainly helps to explain how most insects are able to perform some of their extraordinary aerobatic maneuvers or, indeed, are able to fly at all. But nobody yet understands how the helical flow of air is maintained across the span of the wings.

The discovery of the clap-fling, the clap-fling-ring and the party-streamer mechanisms represents a breakthrough in the understanding of insect flight. There can be no doubt that all three are far more widespread among insects, and probably bats and birds, than has hitherto been observed.

Although insects have been in the air for millions of years and have witnessed the changing scenes of life on this planet before birds or man learned to fly, only now are we beginning to appreciate the amazing complexity of insect flight. Insects have taught us that our understanding of the science of aerodynamics has tended to be anthropocentric in its approach and have led us to the study of the aerodynamics of non-steady airflow. Our knowledge of insect flight is far from complete, and as always, we still have much to learn from nature.

The Feathered Wing

Man has always been more intrigued by the flight of the bird than by that of any other creature. Most bats can be seen only at night, while the rapid and erratic movements of insects baffle the human eye. But the supreme mastery of the air achieved by the bird is plain for all to see. The perfect harmony of its form and function has captivated humans throughout the ages, although the subtle blend of the innumerable elements that make up bird flight eluded scientific analysis until the 20th century. Now that the flight of the dove has at last been interpreted and the vibrations of the hummingbird's wings have been recorded on film, the mystery has, to a large extent, been explained. Fortunately, however, the wonder of bird flight still remains.

Birds are the largest class of vertebrates living today, having diversified into some 8,500 species, compared with approximately 4,000 mammal species. The only other living vertebrate to acquire the power of true flight is the bat, although flying fish, a few rodents, marsupials, one species of frog and a lizard species are capable of gliding in a relatively crude manner. Second only to insects, the highest class of invertebrates, birds are the most biologically successful group of animals that has ever existed and, like the insects, owe their success almost entirely to their ability to fly. Wings and feathers have given birds the power to rise into the air, move rapidly from one place to another, travel long distances, find food not available to other animals, escape from enemies and rear and tend their young in high, safe places.

Of all the classes of animals, birds are the easiest to recognize simply because of their feathers. This characteristic has meant that birds, like mammals, are homeothermic, or uniformly warm-blooded, which has given them a major advantage over most other creatures.

Warm-bloodedness allows an animal to maintain a relatively constant body temperature, usually higher than that of the surrounding air; the blood temperature of birds averages about nine Fahrenheit degrees higher than that of mammals, including humans. By contrast, cold-blooded creatures, such as reptiles, depend entirely on the environment for their body temperature, and although they are lively enough when warm, the cold makes them too lethargic to hunt for food or to escape from their enemies.

Warm-bloodedness in both birds and mammals is accompanied by a four-chambered heart that allows fresh oxygenated blood to circulate throughout the body uncontaminated by "tired" blood, as occurs with fish, amphibians and reptiles. Birds can therefore maintain the

A close-up of the wing feathers of a macaw, FACING PAGE. *Feathers are what distinguish a bird from all other animals.*

Figure 51. Regarded as the first bird, the reptilelike Archaeopteryx *lived 140 million years ago.*

high rate of metabolism so necessary for intense activity, particularly flight, and because they are able to withstand different climates, they can explore all parts of the globe, from the polar regions to the equator.

EVOLUTION

The evolutionary change from cold- to warm-bloodedness is at the heart of the question of how birds first came into existence. It seems difficult to believe that birds, whose movements are freer than those of all other animals, could have descended from what were once thought to be ponderous and cold-blooded dinosaurs; yet there is now widespread support for such an ancestry.

The earliest fossil that bears any resemblance to a bird is *Archaeopteryx* (see Figure 51), unearthed in limestone by workers in Bavaria in 1861, two years after Charles Darwin had shocked the world with the publication of his *Origin of Species*. Up to this time, few people had taken Darwin's theory of evolution seriously, but the discovery of this most famous of missing links provided strong evidence that birds may well have descended from reptiles—or at least from reptile-like animals.

Archaeopteryx lived about 140 million years ago in the Upper Jurassic period, during the age of the "terrible lizards," or dinosaurs, and it bore little resemblance to the birds of today. It had many reptilian features, such as a long, bony jointed tail and claws on its wings, and instead of a beak, it had lizard-like jaws with sharp teeth. As its breastbone had no keel to which powerful wing muscles could be attached, it is generally assumed that the creature was capable only of gliding or, at most, of a feeble flapping. But in other respects, *Archaeopteryx* was like a bird. About the size of a crow, it had birdlike feet and was covered not with scales but with feathers that were structurally indistinguishable from those of present-day birds. The flight feathers were obviously designed for flying, with narrow leading edges and broad trailing edges—a shape that increased their aerodynamic lift-producing qualities.

Because it had feathers, *Archaeopteryx* was undoubtedly a warm-blooded animal and is regarded as the first bird, but its ancestry has been the subject of much ongoing speculation. The general view has been that the thecodonts, a group of meat-eating reptiles of the Triassic period (225 to 195 million years ago), gave rise not only to *Archaeopteryx* but also to pterodactyls and dinosaurs. But there is an awkward gap of some 60 million years between the thecodonts and *Archaeopteryx*, with practically no

fossil evidence bridging the two.

A number of leading paleontologists have rekindled a theory much discussed during the latter part of the 19th century. Thomas Huxley, a champion of Darwin, speculated that birds did not evolve directly from thecodonts but from dinosaurs. Indeed, many paleontologists are now convinced that *Archaeopteryx* was not so much a primitive bird as a feathered dinosaur. There are marked similarities between the skeletons of *Archaeopteryx* and certain small dinosaurs, including one or two examples of a fused collarbone—the wishbone—a distinctive characteristic of birds.

Even more fascinating is the theory advanced by John Ostrom of Yale University in 1969, who suggested that dinosaurs were not cold-blooded reptiles at all but warm-blooded creatures. Support for this theory was given a strong boost in the late 1990s when

The feather structure of today's birds, such as this macaw, is identical to that of Archaeopteryx.

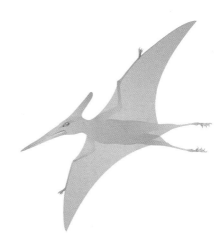

Figure 52. Boosting a wingspan of up to 20 feet, the pterosaur may have glided over mountaintops some 200 million years ago.

a number of other pre-birdlike fossils were unearthed in China —*Caudipteryx* and *Sinosauopteryx* among them. Ostrom now believes that rather than being lumbering pea-brained reptiles, as they have been depicted for years, dinosaurs may well have been intelligent, agile beasts. Scientists Robert Bakker and Peter Galton have gone so far as to substitute dinosaurs for birds in the five broad categories of vertebrates, hitherto classified as mammals, birds, reptiles, amphibians and fish. In other words, dinosaurs are not extinct after all but are living on as birds in our backyards.

It is beyond the scope of this book to enter into this debate, except to say that the theory seems very plausible. As far as the evolution of birds is concerned, the relevant point is that the crucial development of warm-bloodedness may have already taken place in dinosaurs millions of years before they became birds.

At this point, it is worth saying something about the pterosaur (see Figure 52). This extraordinary-looking creature, which evolved from the thecodont, may have been warm-blooded, like the dinosaur. A continuous expanse of membrane connected its hind legs and forelegs in a similar fashion to the wings of a bat. Some pterosaurs had a wingspan that extended to more than 25 feet; clearly, huge muscles would have been essential for flapping flight. As there is no evidence that pterosaurs had such muscles, it seems likely that the large pterodactyls did little more than glide, relying on thermals or rising currents of air around the craggy mountaintops to gain height.

Pterosaurs arrived some time before birds, flourished for over 100 million years and then, quite suddenly in geological terms, became the victim of some unknown catastrophe (together with the massive dinosaurs that dominated the land) and mysteriously vanished some 65 million years ago. With their disappearance, the bird was left to evolve alongside the insect and the bat.

Once small dinosaurs evolved feathers for insulation, it is possible to imagine how flight may have developed. Learning to fly 150 million years ago would have been easier than it is today, because of the denser air at that time, which provided more lift. Running along the ground on their hind legs and leaping into the air in pursuit of the already evolved flying insects, these dinosaurs may have used their feathered forelimbs for shifting the center of gravity to make midcourse corrections (see Figure 53). They may have used their hind limbs as a net for ensnaring insects in flight, just as some bats use their wings and tail to trap insects. It has also been suggested that the most appropriate forelimb movements for controlling roll and pitch in midleap would be similar to rudimentary flapping.

The more orthodox view, however, is that once birds possessed feathers, they evolved flight through gliding down from trees or mountainsides or through launching themselves from branch to branch like today's flying lemurs and squirrels. However they originated, the first wings must have been very crude, functioning as no more than aids to catching prey in midjump or midglide. But by flapping, leaping or gliding, birds' skeletons and general form were modified through natural selection, and birds acquired the ability to fly. The early evolution of flight in birds is comparable to that of insects, in the sense that both may have begun with gliding and then, as new mechanisms were perfected, graduated to flapping flight. Although the early fossil records of birds are incomplete, it is certain that once flight was realized, there was a spectacular species explosion among birds as they rapidly explored new environments and tried new modes of life.

Unfortunately, in the 70 million years immediately following *Archaeopteryx*, only a handful of birds are known to have left any fossil remains. Most of these were seabirds, whose chances of fossilization were much better than those of land birds, the best known being hesperornis, a six-foot-long flightless diving bird with teeth, and ichthyornis, a ternlike bird that had a well-developed keel on its breastbone and was therefore probably a good flier.

The largest prehistoric flying bird ever found was *Argentavis magnificens*. With a wingspan of nearly 26 feet and weighing about 175 pounds, it was well above the theoretical limit for flapping flight and probably functioned only as a glider. Like condors and vultures, it would have relied on soaring on the thermals generated by the Argentinean savanna and would have been able to take off only in a strong headwind. That this gigantic bird managed to survive at all can be attributed solely to the exceptional circumstances at that time.

Fossilized birds become increasingly more common in the Eocene epoch (54 to 38 million years ago). The remains of flamingos, rails and game birds have been discovered in the Paris Basin, and the bones of vultures, herons and kingfishers have been dug up in London clay. One group of fossilized birds found in great numbers was the giant flightless species, such as the extinct elephant birds and moas from the Eocene, Oligocene and Miocene epochs (54 to 7 million years ago). These, together with such extant birds as the ostrich, kiwi, rhea and emu, lacked a keel on the breastbone and so were all considered to represent a stage of development before flight evolved.

At one time, all these birds were believed to be closely related, and so, to distinguish them from the keeled birds, they were placed in a separate subclass known as the *Ratitae*. Today, it is understood that they are not related to one another at all but are the result of evolutionary convergence. Each order of these ratites, as they are now called, developed independently in the part of the world where it is found and probably stopped flying when there was no further need to take to the air to escape from enemies or to obtain food. With disuse, their wings and flight muscles gradually atrophied and their keels disappeared.

Figure 53.
A pre-Archaeopteryx *dinosaur.*

Black-headed gull.

few flightless species left there, such as the kiwi, the takahe and the kakapo, but these populations are barely hanging on by the skin of their teeth and often only on remote islands.

FORM & FUNCTION

Flight imposes strict limitations on the shape, size and structure of flying animals as well as flying machines. If an animal is to fly, aerodynamic efficiency and power have to be combined with structural and muscular strength, and weight must be kept to a minimum. In birds, such requirements have resulted in a certain uniformity of design; for example, a robin has the same basic structure as a pigeon or a duck. On the other hand, animals that live on the ground evolve in many different shapes and sizes (compare man, for instance, with a giraffe or a shrew). Every part of a bird's anatomy is perfectly suited for life in the air. The compact body of the bird combines strength with extreme lightness—a characteristic that is unequaled in terrestrial creatures—while its smooth, sweeping lines offer minimum resistance to the air.

This raises an interesting, and sometimes overlooked, point relating to the startling difference in outward appearance between birds and insects. Insects are generally blunt, complicated creatures with appendages and rough structures sticking out in all directions. Birds, on the other hand, are streamlined, with pointed front ends and smooth, unobstructed contours along their bodies; even their legs almost disappear amid the plumage when in full flight, just as the landing gear of an aircraft retracts.

It's interesting that the loss of the ability to fly occurred in birds on oceanic islands free from predators. Once it has lost this ability, a bird is far less likely to survive changes in its environment, such as the sudden appearance of predators—humans, in particular. Apart from the penguin, which spends most of its time at sea, there are few flightless birds common today. Extinctions over the past few hundred years are relatively much higher in flightless species than in those which are able to fly.

We have only to consider the sad fate of the dodo, the moa and the great auk. And then there is New Zealand, where there were no mammals or other ground-living predators to threaten birds, so many species lost the power of flight and survived peacefully for millions of years—until "civilized" man turned up, complete with guns, traps and introduced mammals. As they were unable to escape, most of the birds became extinct within a matter of years. Fortunately, there are still a

Whereas the majority of insects rarely travel at speeds greater than 10 to 15 miles per hour, birds move at much higher speeds, sometimes exceeding 100 miles per hour. Bearing in mind that the drag at 100 miles per hour is 100 times greater than the drag at 10 miles per hour, it is clear that efficient streamlining is essential if birds are to attain high airspeeds with a minimum expenditure of energy.

The problem that birds faced during their 150-million-year evolution was more than a matter of aerodynamics and wing development. Their whole anatomy and physiology had to be modified not only to combine maximum strength with minimum weight but also to improve the efficiency of bodily functions, such as breathing and blood circulation. Even the senses of birds have been very finely attuned; their eyesight provides the greatest amount of information at the fastest possible speed, reaching a degree of perfection not found in any other animal. In light of such specialized changes and refinements, it is no wonder that humans have failed in their attempts at true muscle-powered flight.

The effects of these changes can be seen all over the bird's body. To reduce weight, the skeleton is remarkably light; the teeth have been replaced by a light horny beak; and even the reproductive organs virtually disappear once the breeding season is over. To accelerate the rate of fuel combustion, the blood temperature has been raised. The metabolism is also helped by an unusually large heart that beats at a fantastic rate—about 600 beats per minute in the robin and 1,000 beats per minute in the hummingbird.

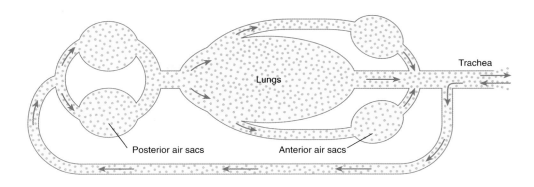

A bird also has a unique way of breathing. In addition to the usual pair of lungs found in all other warm-blooded animals, it has air sacs spread throughout its body within muscles, abdominal organs and even the hollow cavities of its bones. The air sacs are connected to the lungs and windpipe by a complicated arrangement of tubes and shunts (see Figure 54). On inhalation, most of the air bypasses the lungs and flows into a series of posterior air sacs, while the remainder, after passing through the rear of the lungs, flows forward into another set of anterior air sacs. On exhalation, the air flows out of the posterior air sacs through the lungs and out the windpipe, while at the same time, the air in the anterior air sacs flows directly out the windpipe.

Although the essential exchange of oxygen and carbon dioxide takes place in the lungs, the lungs are supplied with oxygen in such a way as to provide a continuous flow into the bloodstream. The system is known as crosscurrent flow and is outstandingly efficient, not only for normal flight, which is strenuous enough, but for high-altitude flying, where the air is thin. Many birds regularly fly at over 5,000 feet on their migra-

Figure 54. This schematic diagram illustrates the principle of crosscurrent flow in a bird's respiratory system. Air enters the windpipe from the right and moves into the lungs via the posterior air sacs.

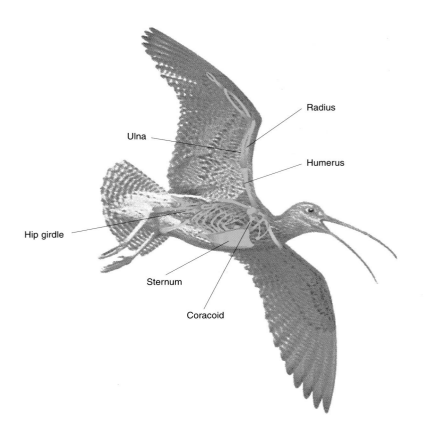

Radius

Ulna

Humerus

Hip girdle

Sternum

Coracoid

*Figure 56*A.
The skeleton of a curlew in flight.

Figure 55. A longitudinal section of the upper arm bone of a bird showing pneumatization.

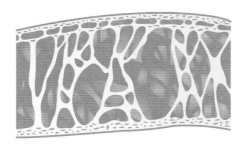

degree the weight of bone present and also strengthening such regions as the skull, chest and pelvis (see Figure 56A).

The hip girdle, for instance, is constructed of a fused plate of bone, so it can withstand the shock of landing and the stresses imposed during takeoff; it also provides a strong surface for the anchorage of leg muscles. The two collarbones, or clavicles, are also fused at their bases to form the wishbone and, together with another bone—the coracoid—help to brace the wings and prevent the body from being squashed when the powerful flight muscles contract. Since extensive neck movement is essential for feeding and preening and as an aid to all-round vision, the neck bones of the bird are markedly different from those of other animals. Whereas nearly all mammals have only 7 neck vertebrae, a bird may have as many as 25, giving it greater flexibility.

Perhaps the most highly modified bone of all is the breastbone, or sternum, which has to anchor the bird's massive flight muscles. For this reason, it has a deep keel jutting out of its center; generally, the deeper the keel, the more powerful the bird's flight. Naturally, the wing bones have also been rigorously adapted to meet the requirements of flight. The largest bone is the humerus, which is short and thick and bears the main flight muscles that are connected at their other end to the keel of the breastbone. The humerus is connected to the forelimb, which is made up of two bones—the ulna and the radius. The inside bone of the forelimb, the ulna, provides a foundation for the secondary flight feathers, while

tions, and some have been plotted on a radar screen at over 20,000 feet. Crosscurrent flow is how they survive flight at such altitudes.

Unlike insects, which are held together by an exoskeleton, the bird, in common with other vertebrates, has bones. Bones not only give the body form and shape but also provide rigid points to which muscles are attached. In the majority of animals, the skeleton is the heaviest part of the body, but many of the bones in a bird are honeycombed with air spaces. Additional strength is imparted by a crisscross of internal bracing struts, just as in the frame of an airplane. This pneumatization, as it is called, is especially apparent in the upper arm bone, or humerus, of gliding and soaring birds (see Figure 55). Several bones that were once separate in the bird's remote ancestors have fused, reducing to some

the wingtip, which is analogous to our hand, comprises a series of finger bones and carries most of the primary feathers largely responsible for propelling the bird through the air. The bone that corresponds to the thumb provides a fixing point for the alula, a small tuft of feathers that reduces the stalling speed.

FEATHERS

Exclusive to the bird, feathers distinguish it from all other creatures. It is thanks to feathers that the bird is so eminently successful, for as well as being soft, warm and colorful, feathers are light and strong yet flexible. Unlike the bat, whose wings consist of skin stretched across bones (of both

forelimbs and hind limbs), the bird's ancestors developed wings from feathers that were stiff enough to form efficient airfoils and take part in active flight. This left the legs free to run, jump, perch, climb, swim, catch prey or, as in the parrot's case, be used as "hands," leaving the forelimbs to get on with the serious matter of flying. By comparison, a bat is awkward and clumsy on the ground.

Feathers make up the entire wing surface supporting the bird in the air. The feathered wing's aerodynamic form, lightness and remarkable ability to change shape represent perfection in functional design. In spite of the saying "as light as a feather," the plumage of

many birds sometimes weighs twice as much as their entire skeleton.

The origin of feathers is still a mystery, yet they have played a crucial role in the bird's evolution. By the time *Archaeopteryx* appeared, feathers were fully developed and have remained essentially unchanged ever since. It is assumed that feathers were derived from reptile scales to form a fluffy insulation and gradually evolved into the highly complex structures possessed by *Archaeopteryx*. Feathers provide insulation in the same way that the hair on mammals does, by holding layers of air between the animal's skin and its surroundings and so helping the body to maintain an even temperature. They also mold the body contours from bill to tail into a smooth, streamlined shape and are important for waterproofing, display and camouflage.

Feathers are of two main types: the down feathers, or plumulae; and the varied outer flight and contour feathers, or pennae. The down feathers lie underneath the cover of the contour feathers, and their chief function is to provide insulation. The beauty and complexity of their structure can be fully appreciated only when viewed under the microscope. The vaned feather is made up of the rachis, a central shaft that is hollow up to about two-thirds of its length. Beyond this point, it becomes solid, to increase the strength at the thinner end. Growing out from the central shaft are hundreds of barbs, which make up the web of the feather. Each barb, in turn, carries hundreds of tiny filaments called barbules, and these are equipped with millions of microscopic hooks, or barbicels, which interlock with the adjacent row of barbules.

The whole structure works much like a zipper and owes its strength and flexibility largely to the barbicels that form firm but mobile links. If the web splits as a result of the hooks becoming disengaged, the bird simply draws the feather through its beak a few times and the condition of the web is restored; this is one of the purposes of preening.

Preening is an essential spare-time activity of birds and, together with periodic molting, provides an effective year-round system of

The contour feathers of a cock pheasant, FACING PAGE.

The flight feather of a parrot, BELOW.

A magnified view of a feather, BOTTOM LEFT.

Figure 56B. Interlocking action of barbules and barbicels, BOTTOM.

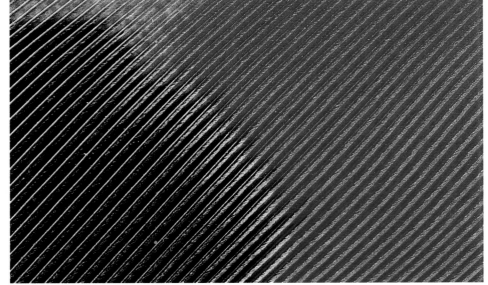

Barb

Barbule

Rachis

Barbicel

The leading edges of an owl's flight feathers, such as this barn owl, have unhooked barbs that create loose fringes which deaden the sound of the wing beats, enabling the owl to fly silently at night.

feather maintenance that is vital to the bird's health and survival. The flight and tail feathers need special attention, because if they become worn or damaged, their aerodynamic efficiency is impaired. But all the rest of the plumage must be kept clean, dry and in the right place.

As some large birds may possess up to 25,000 feathers, restoration can become an endless chore. Chances are that if a bird is not feeding, sleeping, courting or nesting, it is perched in a safe place preening. Feathers have a clear advantage over the chitinous wings of insects or the wing membranes of bats and the extinct pterodactyls: Feather damage is more likely to be confined to a small area and can be repaired, rather than becoming serious or even irreparable. Furthermore, if for some reason the damage is severe, feathers can be replaced by the process of molting and regeneration,

although a bird can sustain a surprising amount of feather damage before becoming grounded.

FLIGHT FUEL

Power for flight is supplied by muscles that function by burning fuel with the help of oxygen carried in the blood. To begin with, the bird's efficient and rapid digestive system breaks food down into suitable carbohydrates and fats; the difference in color between various types of muscles provides a clue as to the nature of the fuel used. Game birds, such as pheasants and grouse, have white flight muscles (as in the "white meat" of the breast) that are fueled by glycogen, a carbohydrate which provides the instant energy requirements for rapid emergency takeoffs and short bursts of strenuous flight necessary to these birds. Because white-muscle fibers operate anaerobically—that is, without oxygen—and the waste products accumulate rapidly, they soon become fatigued and need rest after use. This explains why game birds are unable to summon enough energy for takeoff once they have been flushed out a few times in quick succession. The flight muscles of pigeons and other long-distance fliers, however, burn fat that requires a rich and constant supply of blood to provide sufficient oxygen for extended periods of activity. Consequently, these flight muscles contain a large number of capillaries and so are much darker in appearance. Some muscles, such as those in the breasts of eagles and other soaring birds, are an intermediate color, since they contain both types of muscle tissue and generate energy by burning fat and glycogen.

Only about one-fifth of the fuel is used for working the muscles; the remainder is lost in the form of heat, some of which must be dissipated to prevent damage to the body. A bird circulates blood through its muscles, thereby transferring the heat elsewhere, in the same way that most engines remove excess heat by circulating water through a radiator. Whereas a bat is able to lose excess heat by passing blood through the tiny membranes of its wings, a bird carries it to the lungs, where it is lost through the evaporation of water at the lung surface. If the system becomes overloaded, panting can help to cool it further, or if this is still not sufficient, the bird can evaporate water from the floor of the mouth in what is called "gular flutter."

THE WING

Compared with a fixed-wing aircraft or even insect wings, a bird's wing is an incredibly complicated structure of muscles, tendons, blood vessels and nerve tissue, which, by a combination of muscular movements and feather bending, is capable of flapping and changing its shape in a bewildering variety of ways. Perhaps the greatest advantage birds have over insects and airplanes is the possession of wings that are "elastic" and display variable geometry. A bird can easily adjust the shape and aspect ratio of its wings to suit flight conditions. A stooping peregrine falcon folds the wrist section and primaries to reduce the wing area, minimizing drag (and lift) to gain maximum speed as it hurtles to Earth at 120 miles per hour. But when gently soaring, the bird spreads its wings and tail to full stretch, reducing wing-loading and stalling speed and thereby saving energy.

To bring about this almost infinite range of wing movement, a large number of muscles come into play when the bird takes to the air. The most powerful of these are the two pairs of breast muscles that are responsible for the up-and-down motion of the wings. Between them, these muscles constitute a large proportion of the weight of the body, which in a strong flier, such as the pigeon, may account for more than one-third. Consequently, these muscles are positioned around the bird's center of gravity to help stability of flight.

By far the largest flight muscle is the pectoral, which produces the down beat, the wing's power stroke. The pectoral is anchored to the keel of the sternum so that when the muscle contracts, the wing moves downward. A smaller breast muscle, the supracoracoideus, pulls the wing up again. It is attached to

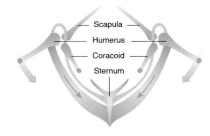

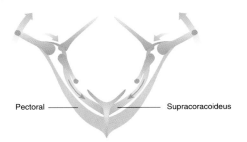

Figure 57. The action of a bird's muscles: top, pectoral muscles contracted; bottom, supracoracoideus muscles contracted.

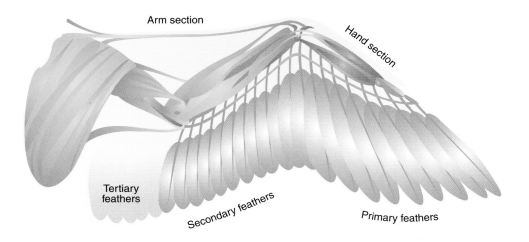

Arm section

Hand section

Tertiary
feathers

Secondary feathers

Primary feathers

Figure 58. The muscles and flight feathers of a bird's wing, ABOVE.

Figure 60A. The two diagrams BELOW *show the effect of the alula in reducing the stalling speed. The wing is stalled, top, while the slot formed by the extended alula restores a smooth airflow, bottom.*

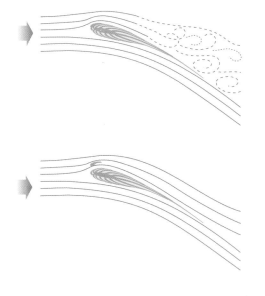

the sternum in a similar way to the pectoral at the lower end, but the tendon at its top is connected to the upper side of the humerus and is looped over the top so that when the supracoracoideus contracts, the wing moves upward (see Figure 57).

Wings, of course, do not simply flap up and down—they also have to be operated and twisted into all sorts of positions about the joints in the shoulder, elbow and wrist (see Figure 58). There are numerous other muscles within the wings to effect and control these movements. Even feather follicles have their own muscles, allowing them to be raised, lowered or moved sideways to assist in flight maneuvers.

The complexity of a bird's wings accounts for one of the major differences between the flight of insects and that of birds. As we have seen, the power and control of insect wings are effected from within the thorax. Whereas insect wings are, for all intents and purposes, flat plates that assume certain airfoil characteristics only when they are in motion, a bird's wings are an example of a perfect airfoil (see Figure 59). The bones in the leading edge

make the wings rigid and rounded, and the feathers at the trailing edge taper to a point. The efficiency of the airfoil is further improved by a hollowing of the wing beneath.

The alula, sometimes referred to as the bastard wing, is an interesting refinement that acts as a subsidiary airfoil in front of the leading edge of the main wing. Under normal flight conditions, the alula is folded back out of the way, but when the airflow over the upper surface becomes turbulent as the bird approaches stalling speed, the alula is spread forward, forming a slot through which air rushes, restoring a smooth, fast airstream and curtailing stalling (see Figure 60A).

The alula functions in the same way as the Handley-Page slot built into some aircraft, and its action can be clearly seen in many of the photographs throughout the book. Slow-flying birds, such as buzzards and crows, make use of the same principle by having deeply slotted primary feathers, so the outer section of the wings becomes a series of narrow airfoils that allow the air to slip through the spaces and thereby reduce turbulence even more.

The broad wings of the sparrow hawk, ABOVE, *are designed for pursuing other birds at high speed amidst vegetation. Here, the alula and talons are extended.*

Figure 59. A section of a wing showing how the feathers make a smooth airfoil outline, RIGHT.

When a small bird like this wren approaches its nest, it splays its primaries to reduce stalling speed.

hours at a stretch. And the larger the bird, the relatively greater amount of power needed to sustain flight. Heavy species, such as swans and condors, have little surplus power beyond that needed for steady flight and find it very difficult to take off or to maneuver with alacrity once airborne. Thus there is an upper weight limit of about 35 pounds for a flying bird.

The relationship between speed and the energy consumed is crucial to birds and aircraft alike. A bird adjusts its flight speed to suit the circumstances. Sometimes, it needs to fly fast when parasitic and form drag become high; at other times, it must fly slowly, which is every bit as strenuous because of induced drag and the extensive flapping required to maintain airflow over the wings. Form drag rises slowly with speed and at lower airspeeds, when wing-flapping becomes more extensive. As it is difficult to calculate, its effect is assumed to remain constant in birds.

Both conditions sap power, but between these two extremes, there is a range of more economical speeds used for normal operational purposes. The relationship is best shown by a power-speed curve (see Figure 60B). The shape of the curve varies between one species and another, depending on weight, wing shape and type of wing beat.

Any technique that helps birds save energy is utilized—in fact, there is a relationship between the amount of time a bird spends in the air and the energy it consumes. Thus a robin, which is airborne for merely half an hour per day, uses over 20 times as much energy flying as it does when quietly perching. A swift, on the other

Whereas soaring birds have almost permanently slotted wingtips, all birds are capable of splaying their outer primaries to a greater or lesser extent to reduce stalling when landing. Insects lack obvious slow-flying devices, but they are able to derive extra lift at slow speed through non-steady airflow effects.

SPEED, POWER, AIR TIME & GLIDING

Since flapping flight is such a strenuous activity, it requires a constant supply of power, sometimes for

hand, which devotes 18 or more hours a day to flying, consumes only a bit more than three times as much energy as when perching. Some birds fly very rarely. Rails and game birds, for instance, prefer to walk about most of the time. Even when hard-pressed, pheasants tend to run rather than take to the air. When they do take to the air, the relative cost in energy, because they are larger and have low-aspect-ratio wings, is far higher than for birds such as robins.

By contrast, birds like the albatross and swift spend most of their lives airborne, as their style of flight allows them to travel long distances to comb vast areas of land or sea in search of food. Flying is economical for these birds because they have evolved to glide or soar with a minimum expenditure of energy and they possess efficient high-aspect-ratio wings. Between these two ends of the spectrum is the majority of species that spend a few hours on the wing each day, chiefly to find food, to look for a mate or to avoid danger.

Occasionally, though, birds seem to fly for reasons that can only be interpreted as pure exuberance and joy. Ravens, rooks, jackdaws and eagles, for instance, have frequently been seen flaunting their powers of aerobatics, including loops, rolls and inverted flying, but such play usually takes place in updrafts or strong winds, so the players do not waste too much energy.

In its simplest form, gliding flight does not require muscle power. Merely by stretching out its wings, a bird flies in much the same way as a fixed-wing aircraft, with the outer part merging with the arm section to form a continu-

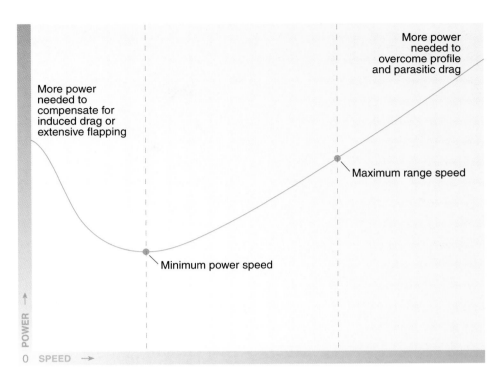

More power needed to compensate for induced drag or extensive flapping

More power needed to overcome profile and parasitic drag

Minimum power speed

Maximum range speed

POWER →

0 SPEED →

*Figure 60*B. *The power-speed curve,* ABOVE. *Most economical flight is between minimum power speed and maximum range speed, depending on the purpose of the flight. Although there is a speed at which power consumption is minimal, outright economy is not always the best approach.*

The slotted primaries of this white-headed vulture, LEFT, *will spread as it slows to land.*

Figure 61.
A raven performs a barrel roll.

Rooks often fly, so it seems, for the sheer fun of it.

ous plane. In this way, many species have so exploited and perfected gliding that they are able to maintain height and travel long distances without flapping their wings at all. Vultures, which have glide ratios of around 10:1, can soar across the African plains with little effort, while the wandering albatross, when aided by suitable winds, is reputed to circle the globe with barely a flap of its wings.

The glide ratio is the ratio between distance traveled and height loss during that time. The larger the glide ratio, the shallower the

glide angle. Among birds, the albatross has the highest-aspect ratio, about 20:1, while state-of-the-art man-made gliders can attain astonishing ratios of about 60:1.

All birds are capable of gliding to a limited extent, even those with low-aspect wings, such as pheasants, once they have propelled themselves into the air with a burst of vigorous flapping. However, small birds, like robins and blackbirds, are poor gliders, because their mass is small compared with their area, causing surface friction (parasitic drag) to retard their forward motion. The matter is compounded by the high induced drag generated by their low-aspect-ratio rounded wings.

By inclining the glide downward, a bird uses the force of gravity to maintain a suitably fast airflow over the wings, just as some insects —or a 747 aircraft, if it must. In perfectly still air, relatively little progress would be made, as most birds would sooner or later reach the ground. But the atmosphere

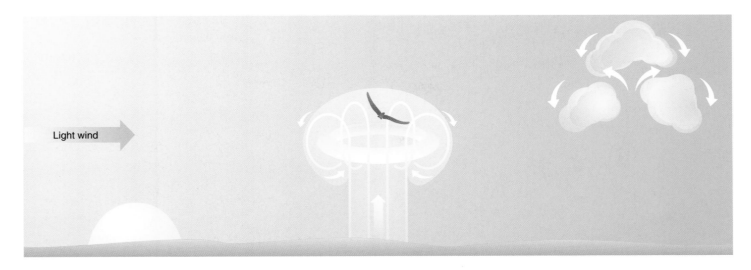

Light wind

seldom remains static, since rising bodies of air are frequently encountered in the form of thermals or moving currents of wind.

Thermals are caused by the uneven heating of landmasses, when rising columns of warm air expand into huge bubbles. As the bubble ascends, currents within it produce a central ring of revolving air, rather like a smoke ring, with a flow of colder air rising through its center (see Figure 62). If the thermal is strong enough, a soaring bird, although flying downward relative to the air, is able to ascend while circling on the updraft within the bubble. The bird must be able to circle within the diameter of the thermal's vortex ring, where the upward airflow is strong enough to sustain its weight.

As the bird turns, some of its lift is applied to another force, centripetal force, which keeps the bird to a circular path and prevents it from sideslipping. The tighter the turn, the greater the loss in lift, which means that the rate of sink increases as the radius of the turn becomes smaller. At the same time, the lower the wing-loading, the lower the sinking speed and the smaller

the radius of turn, so birds with low wing-loading and small size can exploit smaller and weaker thermals than can the heavier and larger species. If the thermal becomes too weak, the bird flaps away to find another thermal that will provide better support.

In this way, lazy birds, like vultures, can ascend in one thermal, then glide or fly to an adjacent one and make long cross-country journeys with a minimum expenditure of energy. Gliding of this kind is called static soaring.

Large broad-winged land birds, such as eagles, buzzards and vultures, are masters of the thermal and are often carried many thousands of feet into the air. Vultures rely almost entirely on these rising bubbles of air for flight and, for this reason, are most commonly found in tropical countries, where thermals are plentiful. In the early morning before the sun has warmed the ground, vultures seldom attempt to fly. The smaller species begin to leave their perches as soon as the weaker thermals start forming, but the larger vultures wait until the sun gets high enough to generate larger thermals before taking off.

Figure 62. The formation of a thermal.

Eagles, storks and vultures are masters of thermal soaring and, like this bald eagle, have long, broad wings with well-spaced primary feathers to improve lift at the tips.

Birds that soar at high altitudes have special characteristics. So that the rate of sink can easily be offset by the rising column of air, they have large, deeply cambered wings of low-loading, well-developed alula and slotted primaries. The wingtips further reduce stalling speed by virtue of their lower angle of attack than that of the arm section—a combination which provides maximum lift and slow-flying capabilities. Extra stability and maneuverability are provided by a large tail.

To us, it may seem the height of ecstasy to soar effortlessly on motionless wings in the hot sun, gazing down on the world below. But to vultures, it is a matter of survival. Carrion feeders need to ascend to heights of several thousand feet with minimum effort in order to survey—with the help of their extraordinarily acute eyesight—a vast area of countryside for suitable food. Soaring enables them to travel long distances by gliding from one thermal to the next and to carry

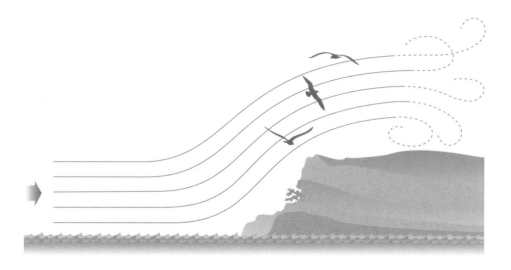

food back to their chicks from up to a hundred miles away.

Many other birds are capable of remaining airborne without using muscle power. Seabirds, in particular, exploit wind currents to maintain or increase height and, like the broad-winged soarers, frequently travel long distances by means of this economical form of locomotion. Seabirds make use of updrafts, but these air currents are of a different nature. They may be caused by various factors, as when an onshore wind hits a cliff, for instance, and is deflected upward (see Figure 63). Even when the wind is offshore, currents of air curl upward as the wind spills over the cliff's edge. Gulls and other seabirds take advantage of these updrafts; fulmars are especially fascinating to watch as they wheel to and fro, sometimes hanging motionless near the cliff tops.

It may seem strange that even the open sea can provide suitable conditions for gliding, or dynamic soaring. When winds blow over

Figure 63.
Updrafts at a cliff face, TOP LEFT.

White pelicans circle together at the bottom of a young thermal, BOTTOM LEFT.

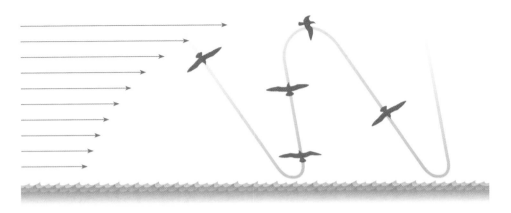

Figure 64. Even the open sea can provide suitable gliding conditions.

water, the lower layers of air are slowed down by friction with the waves (see Figure 64). As a result, a gradient of velocities is produced, with the wind speed at its maximum 100 feet or so above the surface. At 100 feet, seabirds turn to face downwind and enter a dive, gaining airspeed on the way and covering a considerable distance. At sea level, they turn into the wind and use their momentum to climb again into stronger winds. As they do, they progressively lose speed. But this, again, is made up for by the gain in height necessary for the long downwind glide. When they can climb no farther, they turn about, and the cycle is repeated.

The albatross spends most of its life at sea, and many species live in the southern regions of the Pacific Ocean, where the wind sweeps almost continually from west to east. Here, with scarcely a flap of its 10-foot-long wings, this splendid bird covers hundreds of miles. In fact, recent recoveries of ringed specimens prove that they are capable of traveling around the world in 80 days—some 19,000 miles at 40 degrees latitude—by riding the wind. When, on rare occasions, the wind speed drops to zero, the albatross abandons flight altogether and settles on

the water to await the next breeze.

Oceanic birds owe their gliding skill to the special shape of their wings and bodies. Like the static soarers, they have a large wing area, but instead of the heavy build of the vulture family, gulls and albatrosses have slender, streamlined bodies and long, narrow wings to keep form and induced drag to a low level. Such a shape is far more efficient for sustained high-speed gliding, when loss of height must be minimized (see Figure 20A on page 25).

Another phenomenon exploited by oceanic birds and, indeed, by other birds that skim close to the water's surface or the ground is called ground effect, and it occurs when any bird or aircraft flies within about a wingspan of the surface. As the bird gets close to the surface, air is funneled between the wings and the surface, forming a cushion of air; the narrower the gap and the higher the aspect ratio, the more effective the cushion (see photo on page 99).

The result is that induced drag is significantly reduced, which leads to more lift and an extended glide. Birds such as skimmers, shearwaters, petrels, albatrosses, pelicans and cormorants regularly take advantage of this effect, and by flying close to

the waves, Second World War pilots likewise extended their range when short on fuel. Ground effect also benefits large birds, like swans and vultures, that need a takeoff run before reaching flying speed, because it helps them to become airborne before full flying speed is attained. By the same token, it provides extra lift during the last stages of the landing approach.

FLAPPING FLIGHT

In spite of the astonishing gliding ability of some birds, the vast majority of birds can soar or glide for only a few yards, and even soaring birds are unable to do so all the time. Power for flight must therefore be supplied by the bird flapping its wings. Broadly speaking, flapping flight in birds is similar to

that in insects, in that the wings of both have a dual action and work not only as airfoils but also as propellers so that lift and propulsion are effected by a combination of vertical and horizontal movements and spiral twisting. These movements vary according to the species and for different types of flight. For instance, the wing movements of a swan in level flight are obviously not the same as those of a blue tit, while the pattern of wing movement made by a pigeon in fast forward motion is different from that exhibited during takeoff.

Each section of the bird's wing can be visualized as an airfoil that is simultaneously oscillating in the three planes of space around a pivot which is itself moving forward. At the same time, the overall shape of

With its long wings stretched out like a board, a low-flying white pelican takes advantage of ground effect.

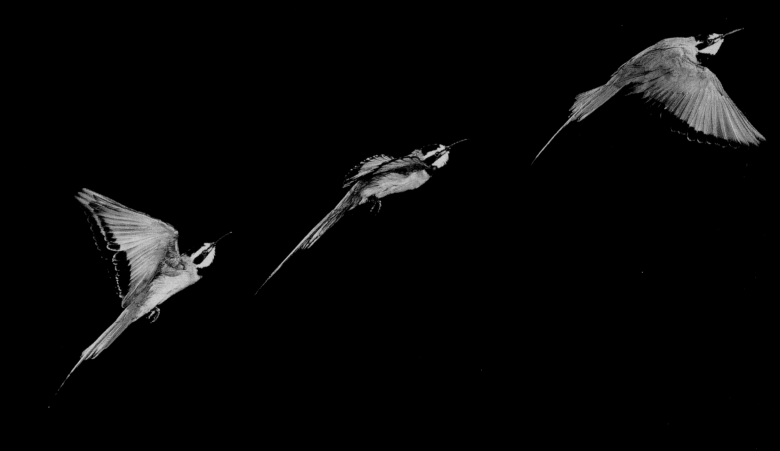

*A multiflash sequence of a bee-eater
some distance after takeoff.*

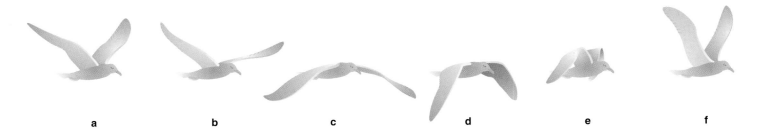

a   b   c   d   e   f

Figure 66. The action of a gull's wings in fast forward flight.

the wing is perpetually changing—a more intricate wing arrangement than anything created by man or any other beast.

During its complete cycle, the wingtip moves farther and faster than any other part of the wing, generally following a figure-eight pattern in a fashion similar to that displayed by insects. The primary feathers provide most of the lift and propulsion, and as they are so flexible, the degree and direction of their bending are a valuable guide to the direction and magnitude of the aerodynamic forces involved. An upward bending, for example, indicates the force of lift, and a forward flexing is one of propulsion. Evidence of these forces at work can be seen in many of the photographs throughout this book.

Functionally, a bird's wing can be considered in two sections: the hand section, bearing the primaries; and the inner part from the forelimb to the shoulder, carrying the secondary and tertiary flight feathers.

In fast forward flight, the movements of the thicker inner part are mainly in the vertical plane and provide a significant proportion of the lift. When the pectoral muscle contracts, the wing moves downward, forcing the air down, which creates an upward thrust on the bird (see Figure 66a-c). At the same time, further lift is generated by the wing's natural airfoil prop-

erties as it moves through the air.

The movements of the outer part of the wing are more extensive and complex, with the wingtip following a tortuous path. During the downstroke, the outer section moves slightly forward as well as downward. In order for the wing to encounter maximum air resistance, the primary feathers overlap to form an airtight surface (see Figure 66c). Thrust is provided at the wingtip by the outer primary feathers, which twist as they are forced through the air.

Each flight feather is so designed that the web is wider and more pliable at the rear edge than at the leading edge. Air pressure created by the wing's movement bends the trailing edge more easily, so each feather resembles the blade of a small propeller. As the primary feathers bend back during the downstroke, the leading edge twists increasingly downward from the base to the tip so that the whole wing becomes propeller-shaped, with the most pronounced twisting toward its weaker end, the tip. In this way, the air is driven backward and the bird is propelled forward.

In level flight, most of the upstroke is one of passive recovery, requiring little effort by the supracoracoideus muscles. As the wing moves up, it is rotated about the shoulder to increase the angle of

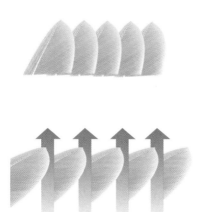

Figure 67. The venetian-blind effect of primary feathers, ABOVE.

A *coal tit* (Parus ater) *streaking from its nest hole at the base of an oak tree,* LEFT, *demonstrates the astonishing wing movements of small birds, especially during takeoff.*

attack. At the same time, the outer wing is partly folded and the primary feathers twist open like a venetian blind, allowing air to pass between the spaces, offering minimum resistance (see Figure 67).

The effect is clearly shown in the multiflash photograph of the little owl on pages 114-115. In a small bird with relatively fast wing beats, the upstroke is largely neutral in its effects. A heavy bird with slow wing beats and higher wing-loading, however, is less able to afford a "wasted" stroke, particularly when flying slowly or taking off. To increase the power of the upstroke, it has relatively large supracoracoideus muscles; to increase the forward drive, the end of its upstroke is usually punctuated by a backward flick of the wingtip. In so doing, a heavy bird derives extra thrust as a result of the air operating against the back surface of the primaries

(see Figure 66e). Certainly, the pigeon adopts this technique in forward flight, as is shown by high-speed photography, and it is reasonable to assume that many other birds must do the same. To what extent is still uncertain.

A small bird with a high wing-beat frequency has more extensive wing movements than a larger species, as can be seen in this coal tit, above, as it streaks from its nest hole.

Slow flight in all birds—as is usually used for takeoff and landing—also demands a much higher degree of wing movement and twisting, reaching a maximum when the bird hovers in still air, thus enabling substantially greater levels of lift and/or propulsion to be generated than would otherwise be possible. The differences between the two types of flight can be seen by comparing Figure 66 with Figure 68.

In slow flight, the bird's body is inclined to the horizontal, the angle

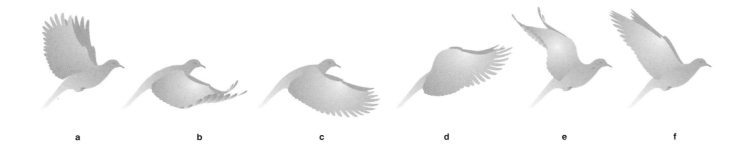

Figure 68. The action of a pigeon's wings in slow flight.

a   b   c   d   e   f

of attack is greater, the alula and primaries are opened and the twist at the wingtip during the downstroke is increased, all of which can be seen in many of the photographs shown here. In both slow and fast flight, the wings are fully extended at the start of the downstroke. In slow flight, the amplitude of movement is far greater, the wings starting their descent at about 90 degrees above the horizontal and sometimes touching at the end of the downstroke (see Figure 68d).

Furthermore, the wings move much farther forward relative to the shoulder, frequently stretching a long way in front of the bird's head. Because of the bird's low airspeed, the inner section of the wing contributes little or no lift or propulsion in normal slow flight; it is the tip section that generates most of the aerodynamic forces.

As soon as the upstroke begins, the upper arm rotates around its axis, throwing the wing into a vertical position, with the tip section

A multiflash of a dove taking off reveals the various phases of the wing-beat cycle. The interval between successive images is more than a cycle.

The kestrel is more often seen hovering into the wind than practicing "true" hovering in static air.

flexed at the joint and the primaries elevated above the bird (see Figures 68e and 68f). At this stage, the whole wing is drawn violently backward and upward and is simultaneously extended. The movement can be compared to the use of the backward flick of the wingtip in fast forward flight, but in slow flight, the movements involved are far more exaggerated, enabling the bird, among other things, to accelerate into the air and so speed up its transition into normal forward flight. There is no sharp division between fast and slow flight; the pattern of wing movement gradually merges

from one to the other. The multi-flash photographs of the owl preparing to land on (see pages 114-115, the dove taking off (see pages 104-105) and the bee-eater (see pages 100-101) illustrate most of these phases.

There is now every reason to suppose that in addition to generating lift by standard aerodynamics, many birds make use of non-steady airflow to provide extra lift, as do insects. As early as 19 B.C., Virgil observed in the *Aeneid* that rock doves made loud claps with their wings when taking off into a steep climb. In his 1890 study of animal locomotion, "Animal Mechanisms," E.J. Marey of France demonstrated that the sounds were caused by the bird actually clapping its wings above its back.

Since then, high-speed photography has shown that when a pigeon takes off in a hurry, its wings are fully extended and almost touching over its body preceding the downstroke. This is followed by a powerful pull of the pectoral muscles, resulting in the wings' rapid downward acceleration. The movement is similar to the insects' clap-fling action (see page 67), particularly as the lifting force generated by this sudden movement is clearly shown by the degree of bending in the primary feathers well before the wings have reached their horizontal position.

Furthermore, calculations made by Torkel Weis-Fogh indicate that under these conditions, a pigeon's wings begin their movement with an air circulation and resultant lift of a magnitude that could otherwise be produced only much later in the downstroke. It therefore seems almost certain that the pigeon—and probably many other birds—makes use of the clap-fling mechanism for generating extra lift when flying at low airspeeds, such as when taking off or hovering.

HOVERING & STEERING

Hovering flight, like slow flight, is strenuous and expensive in terms of energy expenditure and requires pronounced wing movements. Although many birds and bats can hover for a second or two, only a few species are capable of sustained hovering. Hummingbirds use hovering most often while feeding, and kestrels, terns and most kingfishers

A hovering sparkling violet-ear hummingbird. A hummingbird's wing is unique among birds in that the arm bone is very short; the wing is almost all hand, and the joints of the elbow and wrist are locked. When the bird is hovering, the wings sweep forward in the conventional downstroke manner, then rotate on their axis and return inverted. In this way, equal lift is generated during both strokes— power for the upstroke coming from a relatively large supracoracoideus muscle. In the bottom image, the wings are at the start of the downstroke, while the top image shows the wings at the end of the upstroke, after completing one cycle.

hover as a substitute for a perch while hunting for food.

True hovering flight can be defined as flying while remaining in one spot in still air, and under these conditions, the wings generate lift without propulsion. The action of the wings in hovering varies according to the species of bird and its type of flight. One form, which relies on the wind and requires little effort, is practiced by some large birds, such as buzzards and kestrels, which fly or glide into the wind at the same speed as the wind blows them backward, so in relation to the ground, they remain stationary.

The hummingbird is the supreme master of hovering flight—at least in the avian world—although in hovering, it resembles bees and hawk moths more closely than it does birds. Its wing movements are of considerable amplitude, swinging forward and slightly downward in the downstroke and backward and slightly upward in the upstroke. At the same time, the bird's body is inclined at a steep angle to the horizontal, which allows the wings to move through the air in a near-horizontal plane, like the rotor blades of a helicopter. The range of these movements can be seen in the multi-image photograph on page 107.

Indeed, the propeller action of the relatively long primaries is essentially the same as the action of helicopter blades, in that airflow is directed downward, producing lift, rather than backward, as in normal forward flight. A hummingbird can also fly upward, downward, sideways and even backward, its breathtaking ability to maneuver allowing it to dart up to a flower, hover in front of it for a few seconds to sip the nectar, back away and dash nimbly on to the next. Of course, factors such as the scale effect, the absence of profile and parasitic drag at zero airspeed and the unquantifiable assistance generated through nonsteady airflow all play a vital part in the virtuoso performances of this tiny bird.

It is of the utmost importance that a bird is able to maneuver easily in the air; this it can achieve without inherent stability, the two being mutually incompatible (see Chapter 1). Whereas most full-scale aircraft are designed to provide some form of compromise between the two, a bird does not need to be inherently stable, as it can stabilize itself instinctively. When a bird's flight path is disturbed by turbulence, the movement is detected by the semicircular canals in its inner ear; compensations are then made by involuntary adjustments to the wings and tail, and normal flight is resumed. More information about the airflow over and around its body and wings is provided by specialized hairs, known as filoplumes, that are sensitive to air currents. These are distributed around the head and other areas, but little is yet known about their function.

A bird steers in the air by deflecting the airflow to one side or the other, adopting asymmetrical wing positions and tilting the angle of its body or using its tail as a rudder. For straight and level flight, the position and movement of each of its two wings are similar, but when making a turn, the forces are asymmetrical. For instance, a turn to the left is executed by deflecting the airflow to the right while banking to the left and deriving more lift from the right wing. In this way,

the bird is subjected to an external force (centripetal force) acting toward the center of its turning circle, but to avoid inward or outward sideslipping, the force must be related to the bird's weight, airspeed and radius of turn. The centripetal force for the turn is derived from part of the lift, and to compensate for this loss, the angle of attack of the wings during the turn must be increased to obtain sufficient lift. Thus the effective wing-loading during a turn is greater than in level flight, with a corresponding increased risk of stalling, as pilots are only too aware.

The rate at which a bird is able to change direction in the air depends on the relationship between its weight, the airspeed and the area of control surfaces brought into play to produce the turn. At high speeds, only a slight touch to the wings or tail is necessary to effect an abrupt change in direction. Fast fliers, such as swifts, can therefore maneuver easily and do not need

A swallow demonstrates its aerial virtuosity as it darts through a two-inch space in an old stable door.

With the beginning of a powerful down-stroke, a jackdaw kicks off from its nest hole in a church tower.

large control surfaces; hence they possess relatively small tails. On the other hand, slow-flying birds, such as buzzards, have large tails to provide flexibility and stability. A pheasant uses its long tail for twisting between the branches as it bursts from its woodland habitat. A swallow, which is generally fast on the wing but is also capable of

flying slowly in a surprisingly controlled manner, has a forked tail that is closed for high-speed flight but is spread out when the bird flies slowly.

TAKEOFF & LANDING

Takeoff and landing are the two most testing moments in flying, and the skill with which birds effect a graceful and accurate takeoff or landing tends to disguise the inherent difficulties involved in these maneuvers. Both aircraft and birds must make a smooth transition from one medium to another. On takeoff, the basic problem is to gain enough lift to become airborne in as little space as possible, while on landing, it is more a question of maintaining control at low airspeeds.

When a jumbo jet roars down the runway, it seems incredible that the machine will ever leave the ground; but with birds, especially the small species, takeoff appears effortless and almost instantaneous. Indeed, the apparent ease and rapidity with which these creatures rise into the air and attain high speeds is one of the most remarkable features of both bird and insect flight. Some of the disparity between an airliner and a bird can be explained in terms of scale, birds being so much smaller and lighter, but the main difference can be attributed to the higher aerodynamic efficiency of flapping wings at low airspeeds. We know that when birds are in normal fast flight, a substantial part of the lift is obtained from airfoil action. But at takeoff, there is no forward speed, so the initial airflow must be generated by other means, such as running, jumping or the extensive flapping described earlier.

Most birds take off from the ground or a perch by springing into the air using their powerful feet to gain initial impetus. This is followed by the strenuous slow-flight wing-flapping, with the powerful downstroke providing most of the lift and the upstroke most of the thrust (see photo on facing page). As the smaller flight muscles supply the power for the upstroke,

they soon become exhausted, so flight of this strenuous nature cannot be sustained for long periods. Once the bird has built up speed, the pattern of wing movement reverts to normal. The amount of effort sometimes required for take-off is apparent when a pheasant explodes into the air. Its short, broad, deeply cambered wings enable it to extract immense lift in a flurry of intense activity, which is maintained for only a few seconds (see Figure 70).

In spite of its gangly form, the grey heron is also capable of impressive feats of takeoff. Its broad wings help it to negotiate the gaps in forest canopies. I have frequently seen this hardly diminutive bird spiraling up almost vertically from thickly covered woodland streams, its wingtips brushing the leaves during its steep ascent.

Some large birds do not have a special takeoff flight at all but use a technique similar to that of airplanes. Swans, albatrosses and divers, for example, simply turn into the wind and run across the surface of the water or the ground, flapping their wings until they gain sufficient airspeed for liftoff. The hummingbird, rather than pushing off with its feet, creates so much vertical lift with its wings that it pulls the perch up before letting go.

Obviously, the easiest way for a bird to become airborne is by gravity assistance. Seabirds, such as puffins and guillemots, plummet down from cliff ledges, and swallows and martins can launch themselves into the air from the sides of buildings or their nests. Some species, such as swifts, are so built for high-speed flight that this is

Figure 70. The silhouettes of a pheasant, top, and a swift in flight.

A *swift leaves its nest in a roof space*,
ABOVE.

*Like the Concorde, birds must drop their
noses to improve forward visibility
when landing at high angles of attack.
Note that this swan*, RIGHT, *is slowing
down by dragging its tail in the water,
its feet having not yet made contact.*

the only way they can become air-borne—if they become grounded,
they are able to take off again only
with the greatest difficulty. In fact,
swifts are so adapted for a life of
flight that the only time they cease
flying is when they are nesting. All
other activities, including feeding,
collecting nest material, mating and
even sleeping, are performed on
the wing.

Whereas takeoff normally re-quires power and the expenditure
of energy, the transition from air
to ground involves considerable
skill and judgment. In their initial
attempts to land, young birds fre-quently overshoot their perches and
crash into the undergrowth. Birds,
like aircraft, have to reduce speed
for landing to lessen the chances of
injuring themselves. This can be
done in several ways: Large, heavy
birds always try to land into the
wind, but such a technique is not
necessary for the smaller species.
Most birds lose speed by swinging
their bodies into a near-vertical
position and spreading their tails,
which act as air brakes, to create
maximum air resistance (see photo
on facing page).

At the same time, the wings'
angle of attack is progressively in-creased to maintain lift as the air-speed is reduced, while the alula
and slotted wingtips are brought
into play to keep stalling speed as
low as possible. More drastic brak-ing can be effected by flapping in
reverse, just as a jet aircraft uses re-verse thrust to direct the airstream
forward rather than backward.

Birds such as pigeons and wood-peckers lose speed by dipping
down some distance before their
selected landing spot, completing
the final few feet in an upward
glide, thereby reducing the mo-mentum of their flight. The final
impact of landing is absorbed by
the legs in the same way that
an airplane's landing gear does.
Waterbirds, particularly those with
webbed feet, benefit from the
cushioning effect of the water as
they skid to a halt.

Although most waterbirds tend to
land somewhat rapidly, the albatross
has an unusually high stalling speed,

which means that a brisk oncoming wind is a great advantage for slow approaches to its nest. When there is no wind, the albatross has no option but to crash-land, with its feet absorbing the initial impact and its well-padded breast taking the rest as it flops forward.

Once dinosaurs had mastered the physics of flight and evolved into such a diverse range of beautiful creatures, another hundred million years would pass before the next animal managed to launch itself into the skies with any success. Some historians have suggested that without bird flight as a model, man's conquest of the air may well have been delayed for some time, but it is difficult to imagine that such a delay would have been for long.

Returning to its nest with food for its hungry offspring, a swallow uses its tail as an air brake.

A *multiflash photograph of a little owl* (Athene noctua) *coming in to land. As the owl approaches its selected landing spot from normal horizontal flight, it slows down by increasing the angle of attack of its wings and inclining its body to the airstream. At the same time, it spreads its tail, which acts as an air brake, and gradually lowers its undercarriage. Also notice the venetian-blind effect of the wings' primaries for reducing the air resistance during the upstroke.*

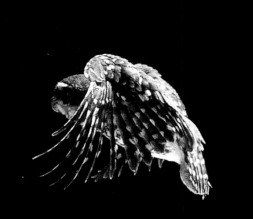

CHAPTER 4

The Evolution of Manned Flight

We should not forget that by the time man started to think seriously about launching himself into the air, animals had already been flying for some 350 million years. Ever since he left his Stone Age cave, man has worshiped the image of the winged object. It appears in the legends, fables and religions of many nations all over the world, but a practical approach to gaining his wings had to wait until man acquired a smattering of scientific knowledge, only a few hundred years ago.

The mythologies and literature of both the Orient and the West abound with winged supernatural personages as well as legendary creatures with wings in the form of dragons, vampires, griffins and jabberwocks. In Egypt, the soul was pictured as a human-headed bird, and in Greece, the psyche was a butterfly; but when evil forces were depicted, their leathery wings were modeled on those of the innocent but nocturnal bat.

Although legends are full of magic carpets and witches on flying broomsticks, there can be little doubt that man's deeply ingrained desire for wings is largely due to his envy of the bird. The bird has been not only his chief source of inspiration to fly but also a model on which his own efforts to do so could be based.

FACT & FOLKLORE

Nobody knows exactly when, where or how the first attempts to convert those dreams into reality were made, for it is here that myth and truth become confused. It is not always possible to distinguish between the actual and the imaginary early flights. Some were unquestionably fanciful enterprises, but perhaps it does not matter so long as the underlying theme upon which all these tales are woven is recognized: the tantalizing lure of the sky and man's constant longing for wings.

As the desire for flight stemmed from an emotional rather than a rational urge, it is not surprising that the early attempts lacked any scientific basis. In any case, systematic scientific thought was still unknown; man had no concept of lift or airfoil action. It seemed that his only line of approach was to try to emulate the bird, which he did, as well as attempt a few other intriguing and bizarre experiments.

The most celebrated of these is encapsulated in the legend of Icarus, whose father, Daedalus, made him a pair of feather wings

Daedalus and Icarus, from a woodcut of 1493.

A medieval tower jumper.

held together by wax. Icarus flew so close to the sun, however, that the wax melted and he hurtled to his death in the sea below, while his less ambitious yet horrified father flew safely across the Aegean Sea to Italy.

One of the earliest—and more likely—endeavors to take to the air was reputed to have occurred about 1500 B.C. when King Kai Kawus of Persia tethered four starving eagles to his light aloe-wood throne; he encouraged the birds to fly by suspending legs of lamb on spears above each corner. The famished eagles are said to have raised the throne a short distance before giving up the hopeless struggle.

In early historical times, the development of aeronautics also remains vague. Aeronautical devices in the form of arrows and boomerangs have been in use since the earliest days of civilization. Strictly speaking, the arrow is a simple projectile with stabilizing feathers, while the boomerang is a power-launched rotating glider. The aerodynamics of the boomerang are so extraordinarily complex that it baffles the imagination as to how primitive tribes ever developed such a device.

Kites, too, are of great antiquity, dating to around 1000 B.C., and yet the mechanics of how they work were not fully understood until the 19th century. Kites originated in China, where they were used not only as toys but also for warfare, and over the centuries, before the development of the airplane, huge man-carrying kites were employed for spying and dropping firebombs over enemy territory. In the 19th century, English schoolmaster George Pocock constructed a large kite for pulling his horseless carriage around the countryside at astonishing speeds, much to the bewilderment of onlookers. Although kites are actually tethered gliders, they were not seriously considered as a potential means of aerial locomotion until the 19th century, and they played little part in the early history of aviation.

ORNITHOPTERS & LEONARDO DA VINCI

There are very few reliable records of man's attempts to fly until early medieval times, when the so-called tower jumpers appeared in force. Among these many courageous but perhaps misguided men was a 12th-century Saracen of Constantinople, who provided himself with a voluminous cloak and leapt off a tower; of course, he crashed to the ground and was killed. In 1503, after affixing wings to his arms, an Italian mathematician named Giovanni Danti jumped from a tower at Perugia; he was seriously injured. Four years later, John Damian, Abbot of Tungland and physician at the Scottish court of James IV, attempted to fly from the battlements of Stirling Castle using wings. Needless to say, he, too, fell to the ground and suffered a broken thigh.

Leonardo da Vinci, in the 15th century, was the first person to give the problem of human flight serious consideration. His genius in this field, as in many others, is unquestionable. Driven by an insatiable curiosity, Leonardo was obsessed with all flying creatures, and birds, in particular, were the objects of his constant observation. His profound understanding of anatomy enabled him to study the structure of a bird's wings and to observe their

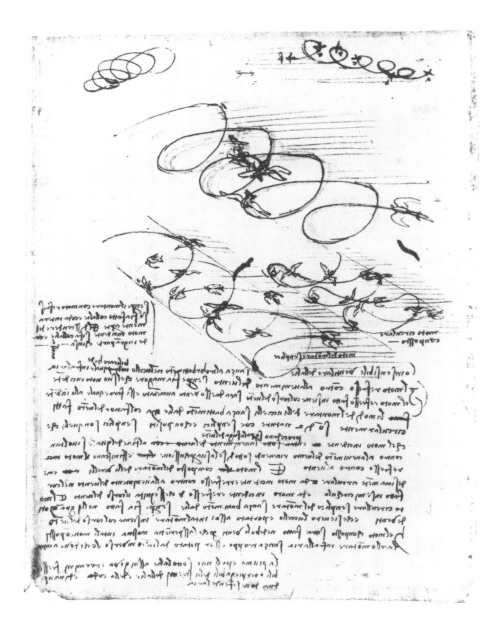

movements both in flapping and in soaring flight. He tried hard to understand the relationship between wing movement and its effect on air currents and investigated pressure, center of gravity and streamlining. He soon realized that bird flight could not be thoroughly understood without a knowledge of the air and its forces.

Leonardo not only recognized the force of inertia, first scientifically expressed by Galileo a hundred years later, but also foreshadowed Newton's third law of motion. He observed that "the movement of the wing against the air is as great as that of the air against the wing." He made copious notes and drawings of all his investigations in the books he kept assiduously for some 40 years, and from these, it is clear that Leonardo was preoccupied with the idea of a mechanical imitation of the flapping flight of birds and bats.

Consequently, much of his work concentrated on aircraft with flapping wings, or ornithopters; but in applying himself to this end, his genius proved misguided. For one

Sketches of bird flight taken from Leonardo da Vinci's notebooks, which were written in mirror writing.

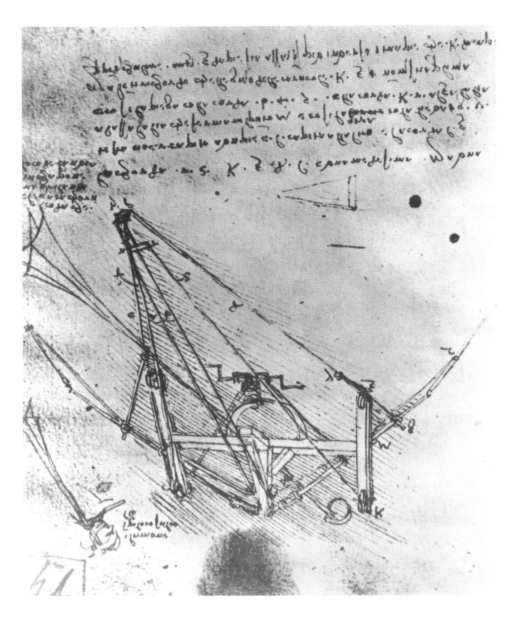

Leonardo's designs for ornithopters showed that he intended them to be powered by a combination of leg and arm movements activated by means of mechanical devices.

tremely intricate and heavy machinery. It is just as well that a full-sized model of one of his contraptions was never tested, since a flight based on backward-flapping wing movements would have proved disastrous for both pilot and machine. Human muscle power, moreover, would have been totally inadequate for propelling Leonardo's cumbersome machines. Despite the current availability of extremely light, strong materials, it seems unlikely that sustained man-powered flights will ever be practical.

He proposed an alternative way of powering his ornithopters, however, which involved a bowstring mechanism that had to be rewound by the pilot in flight. He also suggested a machine with fixed wings, to whose outer ends were attached panels hinged for flapping. Curiously enough, not until after Leonardo had spent all his energies on ornithopters did he write his extraordinary treatise on bird flight, *Sul Volo degli Uccelli* (1505), but in it, he was still unable to discover the secret of bird propulsion.

It is a pity that in all his works, Leonardo clung so tenaciously to the concept of the flapping wing and neglected the fixed-wing glider, which appeared later in his notebooks only in the form of thumbnail sketches. One of these drawings shows a man clinging to a flat board seen in gliding descent, an idea that, if he had developed it, could have led to a vital change in his views of flight and to more fruitful conclusions.

In addition to designing flying machines, Leonardo was the first to consider the principle of the parachute and the helicopter, and he designed a head harness to operate an

thing, he had compared bird flight to swimming and rowing, erroneously deducing that birds "row downwards and backwards." Such a conclusion seems hardly surprising considering the speed at which birds vibrate their wings. Even the slower actions of eagles and gulls were too fast for the accurate analysis that was made possible only by photography.

Leonardo's designs for ornithopters show that he intended them to be powered by a combination of leg and arm movements activating ex-

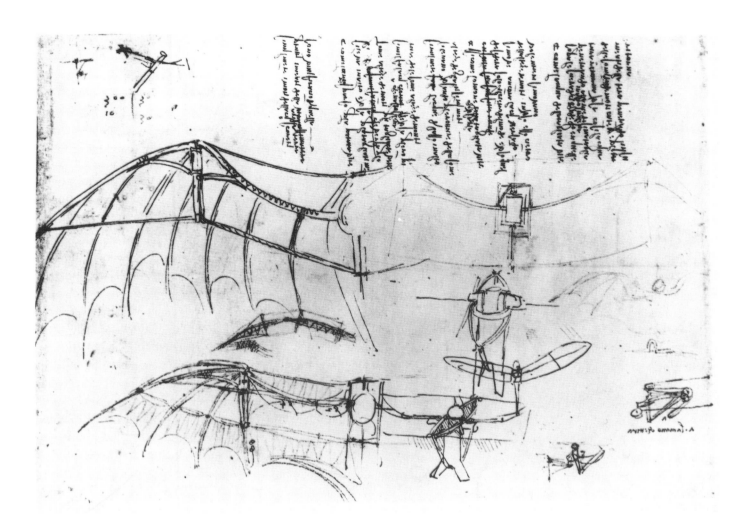

elevator intended to control an aircraft's rate of climb and descent. Yet in spite of his painstaking efforts in recording all his experimentation and research, his work remained virtually unknown for more than 300 years, until the end of the 19th century. Thus, tragically, Leonardo's aeronautics failed to influence the future history of flying. If his research had been published sooner, as he had intended, manned flight might have taken a very different course.

After Leonardo's death in 1519, nearly a century passed before any other thinker gave flying further serious attention, although the tower jumpers continued their frightening escapades well into the 17th century. For instance, in 1663, the Marquis of Worcester published an account of his efforts to make a reluctant 10-year-old boy fly from one end of a barn to the other; the aeronautical details of this event are far from clear. In 1673 in Frankfurt am Main, a brave gentleman by the name of Bernoun broke his neck and both legs when he tried to fly with flapping wings of some sort, and a French tightrope dancer by the name of Allard was seriously injured when he attempted a flying display before Louis XIV.

Unfortunately, the concept of human muscle-powered flight was unwittingly perpetuated by Francis Willughby (1635-72), one of the first British naturalists to treat the study

A later sketch of Leonardo's portrays a gliderlike machine.

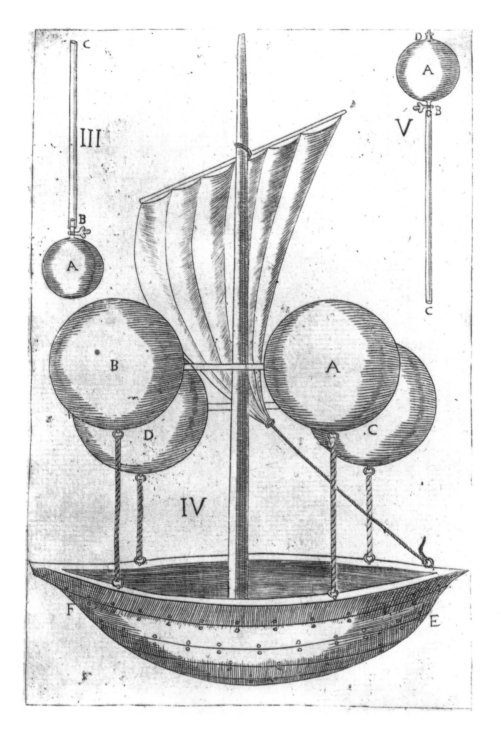

Father Francesco de Lana de Terzi's aerial ship of 1670, taken from a contemporary engraving.

yet more unworkable contraptions based on leg movements.

However, two 17th-century scientists, Robert Hooke (1635-1703) in England and Giovanni Borelli (1608-79) in Italy, quite independently came to the conclusion that human muscle-powered flight was impossible and claimed that unless man could manage to decrease his weight or increase the size of his muscles, flapping flight would be impossible. Today, we understand that it is simply a question of power-to-weight ratio. Man is not designed to fly; he is far too heavily built, with large, solid bones and weak breast muscles.

By contrast, many bird bones contain air, while a bird's breast muscles are enormously powerful compared with those of man. Even if we were to attach highly efficient airfoils to our arms and legs, we would still be incapable of flapping flight. To house suitably powerful muscles for flight, it has been calculated that humans would require shoulders six feet broad. As Hooke observed, an independent source of power was essential.

Despite the deterrent findings of Hooke and Borelli and the ridicule and contempt to which most would-be aviators were subjected, man's irrepressible ambition to fly continued to manifest itself in the form of ornithopters and flapping wings powered by the human frame. Perhaps the most celebrated performance was undertaken in Paris by the Marquis de Bacqueville, who, in 1742, fastened wings to his arms and legs and, cheered on by a huge crowd, leapt from a house in an attempt to fly across the Seine. He floundered miserably and fell onto a washerwoman's

of birds as a science. By calculating the strength of the pectoral muscles of birds and comparing them with the human arm, he concluded that any hopes of flying would depend on the leg, not arm, muscles being harnessed to power the wings. This encouraged the development of

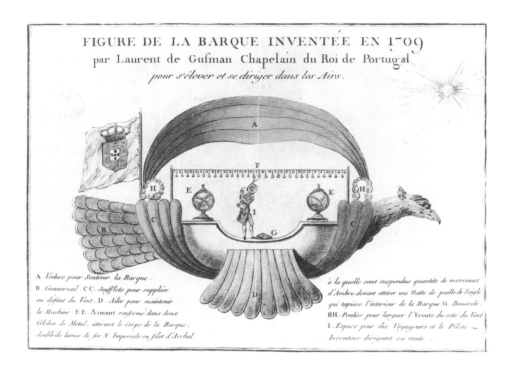

barge, breaking both of his legs.

As an alternative to wings, the 17th and 18th centuries also saw various intriguing and mostly wholly impractical schemes to get man airborne. Father Francesco de Lana de Terzi, a Jesuit priest, published a book in 1670 that included a design for an aerial ship which was to be suspended by four copper spheres emptied of air.

Father Francesco even gave his aerial boat a sail, not knowing that his craft would be unable to obtain advantage from the wind, as a free balloon must inevitably drift with the wind. He also failed to appreciate that the atmospheric pressure would have collapsed his spheres. Nevertheless, his idea represented the first conception of a lighter-than-air craft.

Cyrano de Bergerac (1619-55) suggested an even more ingenious mode of aerial locomotion for lunar travel: Observing that dew rose when sunlight fell on it, he proposed to strap around his body bottles of dew, which, he reasoned, would be soaked up by the sun and hence carry him up too.

One of the most absurd contraptions was the Passarola, the invention of Bartolomeu Laurenço de Gusmão (1686-1724) of Portugal, which was, until recently, treated as a work of pure fantasy. As it appears in prints, it resembled something between a balloon-cum-boat and a bird-powered parachute, but there is reason to suppose that this device may have been tested. Although the Passarola itself could never have flown, evidence now suggests that de Gusmão did build a model which made a tentative flight before the king of Portugal in 1709. A contemporary report strongly indicates that it functioned as a primitive hot-air balloon, insofar as it got off the ground at all. Thus, while Archimedes conceived the principle of flotation, it was de Gusmão who, 2,000 years later, may have been the first to demonstrate lighter-than-air flight.

An engraving illustrating Bartolomeu Laurenço de Gusmão's Passarola, about 1700.

BALLOONS, CAYLEY & THE BIRTH OF MODERN AERONAUTICS

Without doubt, the invention of the balloon during the 18th century had a profound effect on the progress of winged flight. Balloons not only provided a rival method to winged flight for aerial navigation but could also be used as testing rigs for parachutes, propellers and, later, aero engines. It is strange that for thousands of years, scientists had not realized the implications when rising columns of hot gases produced by forest fires and volcanoes carried large fragments to great heights. This discovery was the work of French papermaker Joseph Michel Montgolfier in 1782. While contemplating by the fireside, Montgolfier was struck by the lifting powers of hot air. Without understanding what made the air rise, he devised an experiment in which he lit a fire under a bag made of fine silk. Much to his delight and that of his landlady, the bag filled with hot air and rose to the ceiling.

In June of the following year, Montgolfier and his brother Jacques Étienne demonstrated, for the first time, a large model hot-air balloon before a big crowd in the marketplace of Annonay. Five months later, in Paris, man's first aerial journey took place in the Montgolfier, and shortly afterward, also in France, the first man-carrying hydrogen balloon was released. Thus man floated into the air. It had taken centuries of striving, in which countless men had lost life and limb with flapping wings and other fantastic machines that were fabricated in total ignorance and with a complete disregard for the laws of physics. At last,

man could become airborne at will.

Following the discovery of ballooning, there was curiously little progress or even interest in winged flight until the beginning of the 19th century, when one of the greatest figures in the history of flying appeared on the scene. The invention of the balloon in 1783 fired the imagination of Sir George Cayley, who was 10 years old at the time. From that moment until his death in 1857, Cayley devoted his whole passion and energy to the theory and practice of flying. He was the first man of science to apply his mind to the fundamental principles of mechanical flight, publishing his seminal work, "On Aerial Navigation," in 1809. This paper, in which he prophetically wrote, "Aerial navigation will form the most prominent feature in the progress of civilization," laid the foundations from which all subsequent developments in aviation grew. This and his later works marked the turning point in man's long struggle to acquire his own wings. Historians now consider Cayley to be the true inventor of the airplane.

By reviewing everything that was already known about air pressure and ballistics and by conducting his own investigations into air resistance, Cayley was able to formulate the new science of aerodynamics. Like Leonardo da Vinci, he looked to the bird in the hope that it might provide some clue, and his observations reveal that he was the first man to come to grips with the basic technique of bird flight. He soon discovered that a bird, once airborne, expends far less energy when flying in a straight line than when taking off. Further laboratory

Man's first aerial journey in a Montgolfier hot-air balloon in 1783, as shown in a contemporary engraving.

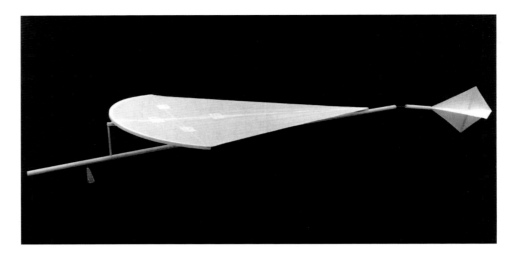

A model of Sir George Cayley's glider based on a kite, 1804.

experiments on the resistance of air to an inclined surface moving through it led Cayley to the perceptive conclusion that a bird in full flight somehow derived its support from the air itself, while its flapping wings were used mainly for propulsion. He was even able to calculate that a gliding rook would be capable of maintaining height when its airspeed reached 25 miles per hour.

Cayley also recognized the potential lifting power of a wing when he discovered that the upper surface had a region of lower pressure than that on the underside. From this fact, he surmised that a cambered airfoil might generate more lift than a flat surface. In his paper "On Aerial Navigation," Cayley summed up the whole problem of mechanical flight thus: "…to make a surface support a given weight by the application of power to the resistance of the air." So, at last, lift and propulsion had been separated, and the way was open to the solution of winged flight by man by the application of power to fixed wings. Borelli had been proved right, and the imitation of flapping birds' wings could be discarded forever.

Most of Cayley's practical work was done at Brompton Hall near

Scarborough, Yorkshire, and it was here, in 1804, that he made the first successful airplane in history. It was a five-foot-long model glider based on the kite; the main plane was fixed, while the tail unit was attached by a movable joint and acted as an elevator and a rudder control.

A movable weight was also provided for adjusting the center of gravity. Cayley had already recognized the importance of stability, and his airplane provided the first practical demonstration of how that could be achieved. In the summer of 1809, when he was 36 years old, he built a full-sized glider which he tested unmanned and which also flew a short distance carrying a boy. Cayley recorded that the glider would frequently lift the boy up and convey him several yards.

After designing a variety of ingenious flying machines of more historical interest than practical value, Cayley returned later in his life to gliders. In 1853, he constructed a full-sized triplane with built-in longitudinal and lateral stability; it also had control in the form of an elevator-cum-rudder operated by the pilot. It was in this machine that his reluctant coachman was persuaded to fly across the valley at Brompton and so make the first gliding flight in history. Full details of this machine's design were published in a popular contemporary magazine. Unbelievably, no one in the scientific world took any notice of it at all.

Cayley, like Leonardo da Vinci, was certainly a genius but was much more practical in his approach to flight. He anticipated in almost every way the airplane of a century later and expressed his firm conviction that "his noble art will soon be brought home to man's convenience and that we will be able to transport ourselves and families and their goods and chattels more securely by air than by water and with a velocity of from 20 to 100 miles per hour." In spite of his accurate predictions and the publication of his brilliant research, however, Cayley and his work were largely ignored until some 20 years after his death—even by the Aeronautical Society of Great Britain, established in London in 1866.

THE RETURN OF THE FLAPPERS

Serious interest in mechanical flight was rare in the middle years of the 19th century. One reason for this was that the secret of flotation (Archimedes' principle) was far simpler to comprehend and apply than the aerodynamic principles governing winged flight. Various other factors combined to postpone exploration along the lines suggested by Cayley, such as the unsuitability of the steam engine and the continuing preoccupation with ornithopters and human muscle power.

Even while Cayley was alive, the subject of winged flight was regarded with skepticism and ridicule by the public, which was hardly

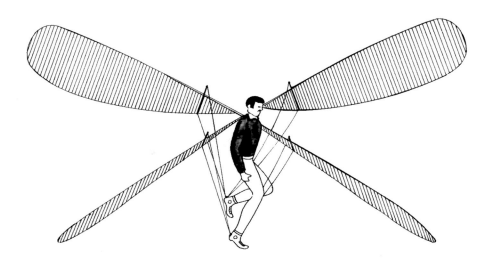

Bourcart's ornithopter, 1863.

surprising, since there were all manner of eccentrics, showmen and charlatans about who did little to add dignity to the science of aviation. The "flappers" still seemed as determined as ever to master the skies with oscillating wings, stubbornly refusing to accept the impossibility of muscle-powered flight.

For example, at the age of 72, General Resnier de Goué of France assembled an ornithopter based on an antiquated design and, in 1788, leapt off the ramparts of Angoulême into the river. Fortunately, he did not injure himself on this occasion, but on making a subsequent attempt, he broke a leg. A clockmaker from Vienna, Jakob Degen (1756-1846), constructed a machine that was half ornithopter and half balloon. The two umbrellalike flapping wings were 130 square feet in area and were ingeniously designed to allow silk bands to open and close in the same way as do the primary feathers of a bird. Assisted by the balloon, which actually supported over half the total weight, and pushing with all his strength, Degen was able to rise off the ground. In spite of using balloons in a number of public demonstrations, he managed

HUMAN MUSCLE-POWERED FLIGHT

Man's goal of self-powered flight was finally realized in 1977, when the Gossamer Condor flew a one-mile figure-eight course. Designed by Paul MacReady, the 60-foot-span aircraft can best be described as a high-aspect-ratio pedal-driven cycle-cum-glider. It was the product of considerable expense and NASA technology, but for all practical purposes, human muscle-powered flight is still impossible—flapping flight, certainly so.

to gain a wide reputation for flying unaided, but in the end, he was attacked and injured by an irate and disappointed crowd in Paris.

An even more bizarre ornithopter was the steam flapper, designed in about 1830 by an Englishman named F.D. Artingstall. Encouraged by the progress of railroad steam engines, Artingstall built a full-sized steam-powered ornithopter. For its first trial, he suspended it from the ceiling, but it flapped so violently that it shook itself to pieces, finally exploding the boiler. Unperturbed by this, Artingstall constructed a second machine even more preposterous than the first—this time with four wings flapping alternately like a gigantic dragonfly. An explosion put an abrupt end to this experiment too.

THE FIXED WING PREVAILS

Fortunately for the future of aviation, a small band of optimistic enthusiasts still believed in the ultimate success of manned flight, and here Cayley's published papers played a major role in guiding these pioneers to victory. One of Cayley's admirers was English engineer William Samuel Henson (1812-88), who described Cayley as the "father of aerial navigation." In 1842, Henson filed a patent for the Aerial Steam Carriage, one of the most remarkable and influential airplanes in history and the first design for a fixed-wing propeller-driven monoplane.

Although the machine was never built, its design was years ahead of its time, owing much to Cayley's influence. It was to have double cambered wings made up of spars and ribs covered with fabric (not to become standard practice until 1908); the wings were then to be braced with wires and king posts. The tailplane had horizontal and vertical movable surfaces to act as elevator and rudder. Henson intended the machine to have a wingspan of 150 feet—huge for those days—and a 30-horsepower steam engine driving two six-bladed propellers.

This extraordinarily prophetic machine incorporated all the principles of aerodynamics then known, except wing dihedral, and many features that would not be developed or tested until much later in the century. The aircraft was to be launched by being sent down a ramp to acquire the necessary initial speed. The chief difficulty that Henson faced was the lack of a suitably light engine of sufficient power—a problem common to all aeronautical engineers for the next 60 years. If a viable power unit had been available, the Aerial Steam Carriage might well have flown.

Undeterred by the lack of power, Henson, with the assistance of his friend John Stringfellow, built a 20-foot scale model of the aircraft to verify the soundness of the basic design. It was fitted with a small steam engine of Henson's design, and the first test was made at Chard, Somerset, in 1845.

The machine was launched down a ramp, but because of the excessive weight of the engine and the inadequate lift of the wings, it was unable to sustain itself, performing what was sarcastically described at the time as a "powered glide." In spite of this disappointment, both men still had confidence in the design, but they lacked the necessary money to develop their ideas or build a full-sized machine.

Consequently, they employed

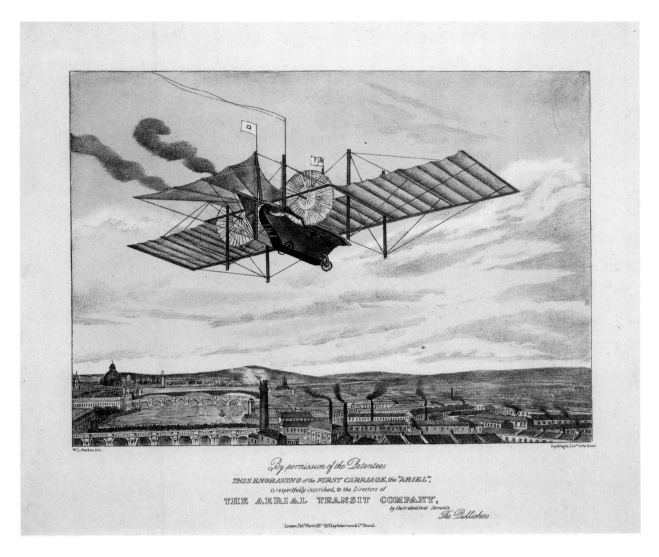

By permission of the Patentees
THIS ENGRAVING of the FIRST CARRIAGE, the "ARIEL",
is respectfully inscribed, to the Directors of
THE AERIAL TRANSIT COMPANY,
by their obedient Servants
The Publishers.

London, Pub.d March 26.th 1843 by Ackermann & Co. Strand.

publicity agents and businessmen to back their enterprise. Unfortunately, the backers got carried away with the project and even tried to promote a bill in Parliament for the formation of an international airline, the Aerial Transit Company, to operate air services to all corners of the globe. One publicity pamphlet was optimistically entitled: "Full particulars of the Aerial Steam Carriage which is intended to convey passengers, troops, and government despatches to China and India in a few days."

Patent drawings, romantic engravings, prints and souvenirs of the Aerial Steam Carriage appeared all over the world, but the enormous publicity had a reverse effect. It succeeded only in heaping ridicule on the overambitious scheme, and the whole project collapsed. By this time, Henson, discouraged by his failures and unaware of the value of his efforts, had immigrated to Texas.

Stringfellow continued to develop Henson's light steam engine and fitted it into various models based on Cayley's multiplane design, but none of them made a sustained free flight. For several years, Stringfellow abandoned aeronautics, but in June 1868, he displayed a model steam-powered aircraft at Crystal Palace, London. It was of

William Samuel Henson's Aerial Steam Carriage flying over a city; the first design for a propeller-driven airplane. From a lithograph by Walton, 1843.

Vincent de Groof's parachute-cum-ornithopter; he was killed trying to fly this machine in 1874.

triplane construction and clearly of Cayley-Henson ancestry but, when tested, proved a failure.

As it turned out, however, this was Stringfellow's greatest contribution to aviation, because it was this machine that persuaded the Wright brothers to adopt superimposed wing configuration in their experiments at the beginning of the 20th century. Although neither Henson nor Stringfellow completed any successful research or offered anything new to the science of aviation, they had shown the world for the first time the shape of the airplane to come, with its fixed wings and propeller propulsion. The Aerial Steam Carriage had fired the imagination of people everywhere, and its impact was felt throughout the rest of the century.

After all the excitement that accompanied Henson and Stringfellow, there followed a period during which a variety of flying machines, both model and full-sized, were constructed, most of which were of indifferent design and had little influence on the future of the airplane. The flappers also still lingered on with their misguided and often suicidal experiments. In 1874, for instance, Vincent de Groof, a Belgian shoemaker, constructed a hand-and-foot-operated parachute-cum-ornithopter, which vaguely resembled one of Leonardo da Vinci's designs. The operator stood within an upright framework, rather like a stepladder, to which the wings and a 20-foot tail were attached. The machine was expected to fly or glide to Earth after being raised from beneath by a balloon, but when cut loose, both wings and tail collapsed from the air pressure, and de Groof plummeted to his death.

A motley assortment of unorthodox ornithopters appeared during the 1860s and 1870s, some powered by muscle and others by steam or even elastic. Still believing that the reaction of a bird's down beat was the total force counteracting its weight and trying to substitute mechanical power for the inadequate muscular energy of the human body, those ornithopterists were further frustrated. They refused to relinquish this notion in spite of the evidence of the almost motionless wings of gliding birds.

Some ornithopters, such as Bourcart's (1863) and Prigent's (1871), were intended to imitate a dragonfly's two pairs of beating wings. Inevitably, they failed, since quite apart from the relationship of power to weight, little enough was then known about the properties of steady airflow, let alone the complex behavior of nonsteady air currents generated by oscillating wings. It seems extraordinary that men were still experimenting with such medieval-looking contraptions only 30 years before the Wright brothers' first successful powered aircraft.

EUROPEAN INITIATIVES

By the latter part of the 19th century, most of the aeronautical initiative had shifted to the continent of Europe, and it was here that the first model airplane took off and landed under its own power. It was built in about 1857 by a French naval officer, Félix du Temple. A clockwork motor was used in early tests but was subsequently replaced with a small steam engine. In 1874, du Temple constructed a full-sized machine of similar design, incorporating wings swept forward and set at a dihedral, a rudder, a tailplane, retractable landing gear and a hot-air or steam-driven propeller. When tested, with a French sailor at the controls, the craft, after taking off down a ramp, leapt a short distance into the air before making an undignified return to the ground.

Ten years later, another short hop was made in Russia in an aircraft designed by Russian engineer Alexander Fedovorich Mozhaiski. It, too, was made to run down an inclined surface before becoming fleetingly airborne, but in view of the assisted takeoffs and the short distances covered, neither of these attempts can be regarded as proper powered flights.

Perhaps the greatest pioneer of French aviation was Alphonse Pénaud, a brilliant figure who committed suicide in 1880 at the age of 30 after his ambitions had been thwarted. He began experimenting with model helicopters but later abandoned them in favor of fixed-wing models that he powered by twisted rubber. One of his machines driven by elastic, the Planophore, deserves particular recognition, because it was the first to be given inherent stability by wingtips set at a dihedral and a tailplane set at a negative angle to the wings. The model made a public appearance in the Tuileries Gardens, Paris, and flew 131 feet.

During his short life, Pénaud pushed the frontiers of aeronautical knowledge forward by establishing both the theory and the practice of stability. But his crowning achievement was an imaginative design for an amphibious monoplane that was patented in 1876 but never built. It was to have cambered elliptical wings set at a dihedral, a rear tail unit with rudder and elevators operated by a single-control column, counterrotating twin propellers, a glass-domed cockpit and retractable landing gear with shock absorbers. The aircraft was even to boast flight instruments, such as a compass and an altimeter. Although the design was years ahead of its time, Pénaud was never to put his farsighted ideas into practice.

For the next quarter-century, the concept of the inherently stable aircraft became an obsession with European aeronautical thinkers. All-round inherent stability is essential for a model airplane if it is to fly without a controller, but in a full-sized machine that has to be flown by a pilot, too much built-in stability makes it difficult to control and maneuver.

During the period that followed Pénaud, the only aircraft to fly were models that relied on their inherent stability, but when these designs were transferred to full-sized machines, this built-in "aversion to change" was mistakenly, though understandably, overdone, creating insensitive and unresponsive aircraft. This would be a stumbling block in Europe until the Wright brothers overcame it some 20 years later in America. The whole problem of stability and control in the air is a complex one involving not only the basic design of the airplane, with its attendant control surfaces and mechanisms, but also the essential human need to master the art of control once airborne.

BIRDS' WINGS AS MODELS

Two other major obstacles to powered winged flight still remained: There was not only the old, frustrating question of power units (lightweight engines of sufficient power were still unattainable) but also the problem of the lifting properties of wings, which in those days were inefficient. In 1866, Francis Wenham's (1824-1908) paper "Aerial locomotion and the laws by which heavy bodies are impelled through the air and sustained" was read before the Aeronautical Society of Great Britain. It was the first significant milestone in aeronautical theory since Cayley.

Wenham's theoretical work was based largely on the observations of bird flight that he made on a voyage up the Nile. While he endorsed many of Cayley's views, he also established that the lift derived from a rapidly moving inclined surface is at its greatest near the leading edge, concluding that a long, narrow wing (i.e., high-aspect ratio) would provide more lifting force than a short, stubby one. Furthermore, Wenham pointed out that all birds' wings are cambered and are at their thickest in front, where maximum lift is derived. "The whole secret," he wrote in his paper, "and success in flight, depends upon a proper concave form of the supporting surface." Before he died at the age of 84, Wenham had witnessed man's conquest of the air.

The theories of the greater lifting power of the cambered wing were followed up later in the century by another Englishman, Horatio Phillips (1845-1924), who provided the scientific basis for all modern airfoil sections. In his laboratory experiments, he constructed and used one of the first-ever wind tunnels and was able to watch the behavior of a jet of steam as it passed over surfaces of different curvatures. Phillips accurately assessed the ratio of lift to drag produced by a wide range of airfoils at various angles of attack. He proved that if a wing is curved more on its upper surface than on its lower, most of the lift generated is the result of the reduced pressure above, rather than the positive pressure beneath, as Cayley had suspected.

However, Phillips was unable to explain this behavior by Bernoulli's principle (see page 22). He imagined that a partial vacuum was created above due to air bouncing off the leading edge. Nevertheless, his work had a powerful influence on the future of aviation.

If one considers the evolution of birds and insects, it seems both natural and logical that the key to human flight lies in the fixed wing. Perhaps if man had been less preoccupied with the search for and application of a prime mover and had been quicker to realize the potential of the motionless wings of gliding and soaring birds, he might have become airborne much sooner.

Not until the latter half of the 19th century, however, were man's efforts redirected, when once again the bird—that unfailing source of inspiration—encouraged him to take the right course. In a book entitled *Du Vol des Oiseaux*, published in France in 1864, Ferdinand d'Esterno (1806-83) suggested that man might try harnessing the wind alone for his motive power. D'Esterno was one of the most influential of those who advocated man's imitation of the soaring flight of birds, remarking, "Whoever has seen large birds of prey sailing on the wind knows that without one flap of their wings, they direct themselves as they choose."

But by reasoning that a bird in flight shifts its center of gravity in some unaccountable way, d'Esterno seems to have misunderstood how the bird effects lateral control. Therefore, when he went on to discuss his design for a glider, he suggested that the pilot control the machine's center of gravity by adjusting his position in a sliding seat. (A similar technique of flight control, whereby the pilot's body was swung about to alter the center of gravity, was adopted by Otto Lilienthal 30 years later and is still used for controlling modern hang gliders and flex-wing microlights.) When d'Esterno's proposals were first published, they were generally ridiculed. By the time they had won a certain acceptance and arrangements had been made for the construction of d'Esterno's glider, he was 77, and he died without seeing his ideas put to the test.

The general skepticism with which d'Esterno's theories were greeted was hardly surprising, considering that the finest avian displays of soaring and gliding were geographically beyond the reach of those most concerned with aeronautics. Although gulls and hawks could be seen in Europe, the true masters of gliding and soaring, such as albatrosses, vultures and kites, are found mainly in the southern hemisphere.

But there were naturally some witnesses to such aerial performances. Charles Darwin, for instance, with his perceptive eye, was so fascinated by the condors of Peru that in his book *The Voyage of the Beagle*, he commented:

"Except when rising from the ground, I do not recollect ever having seen one of these birds flap its wings. Near Lima, I watched several for nearly half an hour, without once taking off my eyes; they moved in large curves, sweeping in circles, descending and ascending without giving a single flap…and the extended wings seemed to form the fulcrum on which the movements of the neck, body and tail acted. If the bird wished to descend, the wings were for a moment collapsed; and when again expanded with an altered inclination, the momentum gained by the rapid descent seemed to urge the bird upward with the even and steady movement of a paper kite. In the case of any bird soaring, its motion must be sufficiently rapid so that the action of the inclined surface of its body on the atmosphere may counterbalance its gravity. The force to keep up the momentum of a body moving in a horizontal plane in the air (in which there is so little friction) cannot be great, and this force is all that is wanted."

In the wake of centuries of bizarre theories and often fatuous experimentation, this extraordinary man was able to summarize the principles of controlled flight after a few hours of bird watching.

The supreme gliding technique of the albatross enchanted French sea captain Jean-Marie Le Bris (1808-72) on his frequent sea voyages to South America. After making careful studies of the bird, he became determined to imitate it, and he set about building a full-sized glider based on the albatross.

The flexible wings were constructed from wood and flannel and were shaped as much as possible like those of the living bird, and the adjustable tail was operated by a foot pedal to steer the machine in the air. For its first test flight in 1857, Le Bris chose to launch his synthetic albatross from a horse-drawn cart. After pointing his cre-ation into the wind and signaling the driver to urge the horse into a gallop, the rope holding the machine down was released, and much to the wonderment of onlookers, the Albatross rose into the air, with its inventor at the controls.

At a height of 300 feet, the triumphant aviator glanced down only to see the driver of the cart swinging below him, entangled in the rope. After a gradual descent, Le Bris managed to make a safe landing for both himself and the terrified man. Although Le Bris made later attempts to become airborne, he was unable to repeat his first success. Quite possibly the weight of the dangling man provided exactly the necessary balance to ensure equilibrium. Le Bris appears to have had no scientific training, but he stands alone in his early struggles to imitate the majestic flight of the magnificent albatross.

Another Frenchman intrigued by the flight of birds was Louis Pierre Mouillard (1834-97), who began his researches in Algeria when he was still a boy. In 1881, he published a book on bird flight, *L'Empire de L'Air*, which became a major source of inspiration to future pioneers of gliding flight. He wrote, "I hold that in the flight of soaring birds (the vultures, the eagles and other birds which fly without flapping), ascension is produced by the skilful use of the force of the wind, and the steering in any direction is the result of skilful maneuvers; so that by a moderate wind a man can, with an airplane, unprovided with any motor whatsoever, rise up into the air and direct himself at will, even against the wind itself." Mouillard believed that wings modeled on the vulture's broad wings

could be made to simulate its soaring flight, but although he built a number of machines, none of his experiments were successful. It was the enthusiasm of his writing, rather than any practical experiments, that was his greatest contribution to the art of flying.

GLIDING FLIGHT

Fired by the fervor of Mouillard's writing, German engineer Otto Lilienthal (1848-96) was one of the most important men in the history of practical flying. Lilienthal was the first to fly consistently long distances in a heavier-than-air craft. He was unquestionably the most successful pioneer of gliding flight in the world, and his knowledge and experience would in turn lead to the success of powered flight.

After examining the work of his predecessors, Lilienthal made up his mind to consider bird flight more intimately than had ever been done before. He concentrated on the aerodynamics of various types of wings and on the relationship between wing area and lift. Although

Cayley had discovered the basic principles of flight, Lilienthal was the first to understand how a bird propels itself through the air by means of the propeller action of its primaries.

In 1889, after 18 years of research and experiment, Lilienthal published *Der Vogelflug als Grundlage der Fliegekunst*, which considered bird flight as the basis of aviation. This book remains one of the great classics of aeronautical literature. Within a year of its publication, Lilienthal became convinced that no amount of theorizing could ever provide him with all he needed to know about the air and that the only way to come to "intimate terms with the wind" was to put his ideas to the practical test. Thus began his first tentative efforts at gliding.

During the next seven years, up to his death in 1896, Lilienthal experimented with various monoplane and biplane gliders. They resembled giant bats, with tough cotton cloth supported on radiating ribs of peeled willow rods. At one stage, he made a glider with twist-

A modern reconstruction of Jean-Marie Le Bris' Albatross.

One of Otto Lilienthal's last flights in his glider in the late 1800s.

ing slots at its wingtips, which were meant to imitate the propeller action of a bird's primaries. The power was to be provided by a small carbonic-acid gas engine, but Lilienthal never attempted any powered flights. His machines were all hang gliders, in which the pilot supported himself by his arms, leaving his hips and legs free to swing in any direction, thereby enabling control in pitch, roll and yaw by shifting the center of gravity.

Lilienthal's first gliding experiments were made in the early 1890s

from a springboard in his garden, but he soon abandoned this in favor of hill-launching. With experience, he learned to use his body to maintain equilibrium and control in variable wind currents and gradually increased both the height of his launches and the distances flown. In five years of intensive gliding activity (1891-96), Lilienthal made some 2,500 glides, with controlled flights of up to 1,150 feet. With the steady improvement in the design of his gliders and in his technique, he planned to devote more time to powered flight by means of orni-thoptering wingtips.

About a year before his death, Lilienthal began to consider alternative ways of controlling the aircraft in flight. In a letter to a friend, he outlined his thoughts on wing-warping, steering air brakes on wingtips and primitive rudder control. In the last few months of his life, he tried to develop a type of body harness for raising and lowering the tail of the gliders, and it was while experimenting with this harness that he had a tragic accident.

On a warm summer's day in 1896, Lilienthal was gliding in his favorite machine when a gust of wind suddenly blew him to a standstill in midair. He threw his body forward in an effort to get the nose down and pick up speed, but it was too late: The machine stalled and plummeted to the ground. He died of his injuries the following day in a Berlin clinic.

Lilienthal's influence on the future of aeronautics was profound. At the time of his death, he was beginning to arrive at a practical method of flight control. Had he lived longer, he would surely have accomplished some form of powered flight before the Wright brothers did. Moreover, the advances made in photography and printing during the last 10 years of Lilienthal's life gave his work an extra dimension. Apart from being more convenient to use, the dry-plate negative—invented by Dr. Richard Maddox in 1871—was fast enough to capture movement. The plate camera had also become a relatively sophisticated instrument, and considerable progress had been made in the production of half-tone printing.

All this meant that Lilienthal's gliding activities were able to be reproduced all over the world in a magnificent series of photographs. For the first time, people could see for themselves that without an engine and using the natural forces of gravity and wind, man could fly. Perhaps the finest tribute to Lilienthal has been the growing band of enthusiasts who, within the last quarter of the 20th century, have been drawn to the exciting and exhilarating sport of hang gliding.

Toward the end of the 19th century, power-driven flight was nearly a reality. In England, as a direct result of Lilienthal's inspiration, Percy Pilcher (1866-99) was carrying out successful glides in machines of his own design and had even constructed an engine for powered flight before his untimely death in a gliding accident. In France, Clément Ader (1841-1925) built a power-driven aircraft that looked like a bat and was reputed to have made a short hop after taking off from level ground. But the immediate future of aviation lay elsewhere. After all the tantalizing slow years of aeronautical progress in Europe, the final stages in the preparation and consummation of man's first sustained powered flight took place on the other side of the Atlantic.

AMERICA & FINAL SUCCESS

Samuel Pierpont Langley (1834-1906), an American engineer, scientist and astronomer, had already flown models of steam- and gasoline-driven aircraft for up to three-quarters of a mile before turning his attention to full-sized machines. In 1898, he was asked by the U.S. government to develop aircraft for military purposes, but when, in due course, his man-carrying airplane was tested, it proved disastrous. On each of its two trials over the Potomac River, it fouled its catapult launching mechanism and fell into the water. Even if the launch had been successful, it is highly unlikely that the aircraft would have flown, since the volunteer pilot was totally inexperienced —he had never flown in a glider, let alone in a completely untested powered machine.

What Langley failed to understand was that successful flight depends as much on the pilot as on the qualities of the aircraft itself. In view of this failure, the war department lost interest in him and withdrew its support, leaving him heartbroken. Because of his misconceptions about flying, Langley made little impression on the science of aviation, but his enthusiasm and confidence in the future of powered airplanes played an important part in persuading the Wright brothers to take up flying.

An even more outstanding aeronautical figure in America was French-born civil engineer Octave Chanute (1832-1910). His interest in aviation began well before Lilienthal's gliding experiments in Germany, and in his hometown of Chicago, he carried out an exhaustive investigation into heavier-than-air flight, collecting every available item of information. Chanute published a series of articles in 1894 that were later reprinted as a book called *Progress in Flying Machines*, which would inspire the Wright brothers to start their own experiments with heavier-than-air flight.

Chanute thought that the underlying cause of the growing number of failures in the air, culminating in the deaths of both Lilienthal and Pilcher, was the lack of stability. He therefore set about designing a glider with a high degree of stability, in which the equilibrium was automatically regained after being upset by gusty-wind conditions. Although too old to fly himself, Chanute rejected experiments with models, believing them to be inconsistent in open air and "unable to relate to the vicissitudes which they encountered."

For these reasons, Chanute secured the help of a young engineer and conducted tests using full-sized gliders. His most successful machine was a biplane, which was a direct descendant of Stringfellow's triplane model that was displayed at Crystal Palace, London, in 1868 and which, in some respects, resembled Lilienthal's biplane hang gliders. It was structurally far superior to anything yet made and certainly helped the Wright brothers to adopt a biplane construction of similar configuration and rigging.

An enthusiastic collector and disseminator of aeronautical information on both sides of the Atlantic, Chanute also took great pains to persuade others to experiment with building and flying airplanes. His close friendship with the Wright brothers undoubtedly did much to encourage and support them in their final conquest of the air.

It seems hard to believe that during all this intense activity with heavier-than-air flying machines, at a time when man was at the threshold of success, balloons and airships were still the main center of aeronautical interest. Both the general public and the so-called experts in the field still considered the airplane a long way off. Even the distinguished scientist Lord Kelvin declared in 1896, "I have not the smallest molecule of faith in aerial navigation other than ballooning." He could not know how rapidly events were to prove him wrong.

Wilbur (1867-1912) and Orville (1871-1948) Wright had been interested in flight since childhood, but their serious study of aeronautics did not begin until after Lilienthal's death in 1896. Until then, the brothers had been engaged in newspaper production and the manufacture of bicycles in Dayton, Ohio, and the money and skills acquired from their businesses enabled them to start on their new venture.

In the early stages, the subject of flight was just a hobby to the Wright brothers, but they soon became totally absorbed. They read all the available literature, including the works of Mouillard, Lilienthal and Chanute, and before long, they arrived at some important new conclusions. They recognized that there were two different prevailing attitudes toward aviation: the first arising from those who made model airplanes and the second from those who actually took to the air themselves. In considering their models as functioning independently, the model makers tended to neglect questions of control and maneuverability in favor of an overemphasis on inherent stability (in those days, there was no radio control, so model aircraft had to be built with much inherent stability; see Chapter 1).

This approach suited model aircraft, for which complete stability was a prerequisite of flight, but held back man's own attempts to become airborne. The aviators, on the other hand, designed and flew

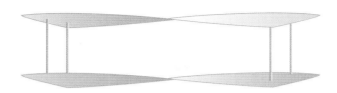

Figure 71. Wing-warping—the wings are twisted to provide lateral control.

their airplanes, with themselves functioning as the central control. They regarded the airplane as a means of learning about and experiencing flying, but unfortunately, they put their lives at continual risk, as both Lilienthal and Pilcher demonstrated. The Wrights, therefore, were determined to avoid the mistakes of the one by finding more efficient ways of maintaining control in the air and the dangers of the other by gaining flying experience without loss of life.

Like many of their predecessors, the Wrights watched birds to see how they managed the fluctuating air currents. Wilbur's observations of the soaring flight of the turkey vulture provided him with a clue. He noted that when the bird was laterally displaced by a gust of wind, it righted itself by a torsion, or helical twisting, of its wingtips; if the leading edge of one tip was twisted downward, that of the opposite wing had an upward turn.

If the wings of an aircraft could be made to twist in a similar manner, the Wrights theorized, the pilot could control and stabilize the aircraft thus, rather than by shifting his body. To achieve this, the wings would have to be light enough to be "warped," as they described it, and yet strong enough to lift the aircraft in the normal way (see Figure 71).

After experimenting with wing-warping on a kite in 1900, the Wright brothers set about making their first glider. As well as incorpo-

Soaring turkey vultures like these gave the Wright brothers the clue to lateral control in aircraft. These birds are ascending on a thermal.

rating a wing-warping mechanism for control in a roll, they added another feature that they used in all their early aircraft: They deliberately made the machine unstable. In designing an aircraft that did not have an inherent tendency to right itself when displaced from straight and level flight, they intended to leave the stability to the skill of the pilot. If a gust of wind caught one wing, causing the aircraft to bank to one side, the pilot would have to level the wings by applying wing-warp; but they made a mistake in thinking that the machine could be steered by means of wing-warping alone.

The glider was completed in the autumn of 1900. It was a biplane with a 17-foot wingspan, the basic biplane concept having originated with Cayley, Stringfellow and Chanute. It had a forward-controllable elevator, and the pilot lay prone to reduce air resistance and minimize the chance of injury to himself while landing. In fact, only a few piloted glides were made in this machine, since it was usually flown as a kite, with the controls operated from the ground.

The Wrights decided to tether the aircraft during its early trials to gain experience from prolonged flying time. The tests were conducted on the lonely sand dunes around Kitty Hawk, on the coast of North Carolina, an area of strong and constant winds. The trials were a great success; not only did the wing-warping device and elevator respond well, but the glider confirmed their belief in avoiding too much inherent instability. The Wrights returned home elated and more determined than ever to make powered flights. "When once a machine is under proper control under all conditions," wrote Wilbur in his diaries, "the motor problem will be quickly solved."

The following year, they constructed a new glider (No. 2), which had several modifications, including an increased wingspan of 22 feet and an anhedral "droop." This time, it was launched by two men and operated as a glider. It completed glides of up to 389 feet, but unfortunately, it had an alarming tendency to slew around and sideslip in the direction of positive warp. The Wrights also found that the wing camber was too deep, producing an excessive movement in the center of pressure as the angle of attack was altered.

Although some successful glides were made, the brothers were far less pleased than before. They now began to doubt the accuracy of Lilienthal's calculations. "Having set out with absolute faith in the existing scientific data," declared Wilbur, "we were driven to doubt one thing after another, till finally, after two years of experiment, we cast it aside and decided to rely entirely upon our own investigations."

In 1902, the Wrights returned to the drawing board, conducting exhaustive research and experiments on all the aerodynamic problems, including the testing of wings in a wind tunnel. The outcome was renewed confidence and their No. 3 glider. Like the No. 2 machine, it was a biplane with the same wing-warping system and anhedral droop but a different camber.

The most important difference, however, was the addition of double fixed fins at the rear. The purpose of the fins was to counteract the previous glider's alarming habit of sideslipping when warp was ap-

plied. In tests, the new machine reacted well in straightforward glides, but serious trouble arose while banking. The pilot's efforts to limit the bank by applying positive warp resulted in the wings dropping farther or swinging backward and causing the aircraft to begin a spin.

Before long, the Wrights recognized that the central trouble was warp-drag (now equivalent to aileron drag), which produced an increase in the resistance of the wings on one side and a decrease on the other. In the act of banking, the aircraft was slipping sideways through the air in the direction of the bank. The resulting airflow on the side of the fixed vertical fins caused them to act as a lever and rotate the wings about their vertical axis. The problem was finally solved by replacing the vertical fins at the rear with a single movable rudder, with its control cables coupled to the warping mechanism. The warping of the wingtips was then automatically combined with the rudder so that the rudder always moved in the direction of the bank, thus counteracting warp-drag.

In this way, lateral balance could be achieved whenever the aircraft was displaced from the horizontal by the wind or by the pilot's intentional banking. The new machine made perfect controlled glides in winds of up to 35 miles per hour and was able at last to perform smooth banked turns, remaining sensitive and responsive to the lightest touch of the controls. "The flights of 1902," wrote Orville, "demonstrated the efficiency of our system of control for both longitudinal and lateral stability." The secret of the Wrights' success lay in their concept of stability and in

Orville Wright's historic first flight in the Flyer at Kitty Hawk, North Carolina, on December 17, 1903.

the combination of wing-warping and rudder.

During the autumn of 1902, the two brothers made nearly 1,000 glides at Kitty Hawk, becoming experienced and skilled pilots in a comparatively short time. At the end of October, they returned home in high spirits and immediately started to draw up plans for a larger machine, this time to be propelled by mechanical power. The major obstacle was the engine, as suitable ones simply did not exist. Not only did these two remarkable men design their own power unit, but they partly constructed it. It had four cylinders, was water-cooled, weighed 200 pounds and produced about 12 horsepower. Since there was no published information available on propeller design, they researched and developed their own propeller.

The Wrights' first powered airplane, the Flyer, was constructed during the summer of 1903. In appearance and control, it was similar to their last glider, except the rudder and elevators were now double structures. With their new machine, the brothers made the now familiar journey to Kitty Hawk. Then came weeks of meticulous preparation and gliding practice with their old No. 3 glider.

Finally, on December 14, 1903, after many mechanical breakages and exasperating setbacks, the Flyer was ready for its maiden flight. Friends from a nearby coastal lifesaving station stood by as witnesses, and the brothers tossed a coin to see who should be the first to fly. Wilbur won. The Flyer ran along a takeoff rail, climbed steeply, stalled and plowed into the sand. Wilbur had put on too much elevator, but luckily, the damage was only slight, and the aircraft was repaired in a couple of days.

On Thursday morning, December 17, the weather was perfect, with a wind of about 25 miles per hour. It was Orville's turn. The engine was run up and, at 10:35 a.m., the restraining rope released. The Flyer gained speed and rose into the air. It flew for 120 feet and landed safely. The flight had lasted for only 12 seconds, but it was, to quote Orville, "the first in the history of the world in which a machine carrying a man had raised itself by its own power into the air in full flight, had sailed forward without reduction of speed and had finally landed at a point as high as that from which it started."

Man had learned to fly by progressing through stages analogous to those of the insect and the bird millions of years before. He had started with short glides, which gradually led to longer glides and then to flight by powered mechanisms refined over the years. Throughout the whole process, the bird had been man's perfect model and constant inspiration. When Wilbur and Orville Wright were studying the principles of flight, they had frequently returned to the bird to check their theories.

"Learning the secret of flight from a bird," wrote Orville, "was a good deal like learning the secret of magic from a magician. After you once know the trick and know what to look for, you see things that you did not notice when you did not know exactly what to look for." By observing the flight of birds, man had eventually found the secret that opened a new era in the history of the world.

The 20th Century and Beyond

The day the Wright brothers took to the air at Kitty Hawk in 1903 represented the turning point in the history of aviation. Curiously, it would be another six years before anyone in Europe believed that controlled flight was actually possible. In those intervening years, Europeans had no real conception of what manned flight was about; they approached the "aeroplane" more as a kind of winged automobile to be driven rather than piloted in an open sky. They were still preoccupied with the idea of inherent stability, which kept them from concentrating on such essential matters as flight control.

The delay is especially surprising when one considers that the principles of flight were largely formulated in Europe by men such as George Cayley, Alphonse Pénaud and Otto Lilienthal. Indeed, without the work of these imaginative pioneers, the Wright brothers would never have made their historic flight.

On August 8, 1908, Wilbur Wright gave a public demonstration of his aircraft near Le Mans, France. The skeptical crowd of pilots and technicians that had gathered to watch stood in stunned amazement as Wilbur put his aircraft through an elegant display of turns, banks and circles with complete mastery and control. While the immediate impact of this flight on the spectators was profound, its effect on aeronautics had even more far-reaching consequences: From that moment, European aviation sprang to life. A year later, Louis Blériot crossed the English Channel, and the airplane became accepted as the world's newest means of travel.

Boosted by two World Wars and spurred on by commercial and military interests, progress in heavier-than-air flight advanced by leaps and bounds. It is difficult to believe that just over 60 years after that first exhibition in France, man reached the moon, yet it had taken him the better part of a million years to learn the secret of flight. In his attempts to fly, he had struggled with all kinds of inventions, from bottles of dew to steam flappers, before he finally met with success. Once a satisfactory formula was found, however, it was used as the basis for the aircraft we know today.

STRUCTURE & CONTROL

In spite of the multiplicity of sizes, shapes and designs of conventional modern aircraft, the fundamental configuration has changed remarkably little over the years. Common to all are: one or sometimes two

The EA9 Optimist, a traditionally designed sailplane with state-of-the-art construction.

Figure 74. The three axes of movement and control surfaces: the longitudinal, the lateral and the normal, or vertical.

pairs of wings supporting a fuse-lage; a tail unit, normally at the rear; landing gear; and an engine. Although gliders and sailplanes are not power-driven, the great majority of aircraft have a power unit to provide the driving force. Whether driven by jet or propeller, all pow-ered aircraft, excluding the few powered by rockets, depend on ac-celerating a mass of air backward, thereby gaining a reaction, or thrust, in a forward direction.

Whereas all animal flight relies on the propeller action of wings to provide thrust, the airplane is de-pendent on some kind of engine in which fuel is burned. The heat pro-duced is then converted into me-chanical work in propelling the air-plane forward against drag. The piston engine has been used for pro-pelling aircraft since the very first powered flights, and in spite of its mechanical complexity, it is still the most efficient engine for slow flight.

The simplest type of engine is the jet, or gas turbine, and although there are various sorts, the princi-ples of their operation are broadly similar. At the front of the engine, air is compressed into a combustion chamber by a compressor, where

it is mixed with fuel and burned. With the energy thus produced, a jet of hot air and burned fuel is ac-celerated out of the rear nozzle at high velocity, producing the forward thrust. Some of the energy released by combustion is absorbed by a tur-bine wheel, which drives the com-pressor at the front.

Most modern subsonic jet air-craft, however, now use the more efficient and quieter high-bypass, or turbofan, engines, in which a large-diameter fan mounted in front of the engine draws a greater mass of air into the compressor; at the same time, a proportion of the air passes around the outside of the engine. Turbofans produce more thrust and less noise and help to keep the en-gine cool. Aircraft powered by pro-pellers, on the other hand, rely on a relatively large-diameter jet of un-heated air being driven backward at a relatively low velocity, the blades functioning like rotating wings. The power for the propeller may come from a piston engine or a gas turbine, in which case it is known as a turboprop.

All aircraft must be controlled in the air along three different axes, fixed relative to the aircraft. All three pass through the aircraft's cen-ter of gravity at right angles to one another (see Figure 74). The longi-tudinal axis runs from nose to tail; the lateral axis runs parallel to a line across the span of the wings; and the normal, or vertical, axis is at right angles to the other two. Thus the normal axis is vertical to the ground only when the aircraft is in straight and level flight. Movement about the lateral axis is known as pitching; about the longitudinal axis, it is referred to as rolling; and about the normal axis, it is called yawing.

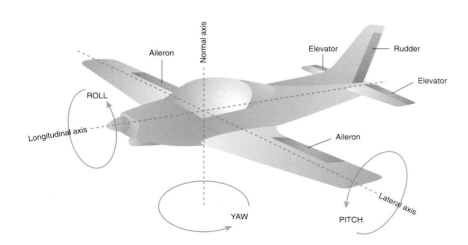

The Beech 2000A Starship 1 is an elegant example of canard configuration.

Whereas a bird is infinitely flexible in the adjustments it can make to its wings and tail to control its flight, most airplanes are restricted to the use of three principal control surfaces: elevators, ailerons and rudder. All three are operated from the cockpit by the pilot—the elevators and ailerons by the stick, or control column, and the rudder by the pedals, or rudder bar, on the floor.

Most important are the elevators, which are normally hinged at the trailing edge of the tailplane, or stabilizer, and effect longitudinal control along the pitching axis. When the stick is moved forward, the elevators hinge downward, causing the air on their upper surfaces to flow faster; the pressure above then decreases, resulting in the tailplane being lifted upward while the nose pitches downward (see Figure 76). When the stick is pulled back, the reverse action takes place: The tail pitches down, and the nose lifts up, with a consequent increase in the wings' angle of attack, which increases lift.

Ailerons have now replaced the warping system originally devised by the Wright brothers for controlling movement along the rolling axis. They are hinged at the trailing edge, usually toward the tip of each wing. When the stick is moved to the left, the aileron on the left-hand wing moves up, the one on the right-hand wing moves down (see Figure 77). Owing to the increase in lift on one side and the decrease on the other, the airplane responds by banking to the left. The reverse action occurs when the stick is moved

CANARD DESIGNS
Although the vast majority of aircraft have conventional tailplanes and elevators, an alternative way of maintaining stability and control in a pitching plane is by using a canard at the front of the machine (either fixed with elevators or all moving). The Wright brothers' Flyer was an example, as are some supersonic fighters; otherwise, canard-configured aircraft are quite rare. An advantage of canard configuration is that an aircraft can be made stallproof. By setting the canard at a higher angle of attack than the wings, the canard always stalls first, thus lowering the nose and restoring normal airflow to both surfaces (see photo above and on page 161).

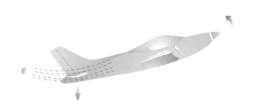

Figure 76. The elevators control an aircraft's pitch.

Figure 77. Ailerons control movement along the rolling axis.

to the right. The rudder controls movement along the yawing axis. By depressing either the left or the right pedal, the rudder will hinge accordingly to the right or the left, causing the aircraft's nose to follow in the same direction (see Figure 78).

As in birds and insects, the movement of these control surfaces at any given angle has a far greater effect at high airspeeds than at low. The area of the surfaces and their distance from the axis about which the airplane turns also influence their effectiveness.

These three basic movements are known as the primary effects of controls, but they have important and interrelated secondary effects as well. For example, in adjusting the wings' angle of attack, the elevators increase or decrease the drag; because of this, the aircraft's flying speed is also controlled by the elevators. At the same time, by chang-

ing the wings' angle of attack, the elevators control ascent or descent.

The application of ailerons and rudder likewise produce secondary effects. Ailerons, which are primarily used to control the bank, or roll, induce what is known as sideslip. If the pilot banks the aircraft to the left by moving the stick to the left, for example, weight and lift become out of line, and the machine will tend to sideslip in the direction of the bank. The fuselage and fin on the left-hand side are then subjected to a greater airflow. As a result, the aircraft behaves like a weathercock and yaws to the left, in spite of the rudder's being held central.

If no action is taken to correct this, a counterclockwise spiral dive will develop. The secondary effect of applying the rudder is to cause the aircraft to roll. When the left rudder is applied, for example, the aircraft will yaw to the left, causing the right wing to move faster than the left (see Figure 79), and because the wing with the faster airflow has the greater lift, the machine will bank to the left, in spite of the stick's remaining central.

Figure 78. The rudder bar in the cockpit controls the rudder and movement along the yawing axis. To execute a balanced turn, the pilot applies aileron and rudder together in the same action.

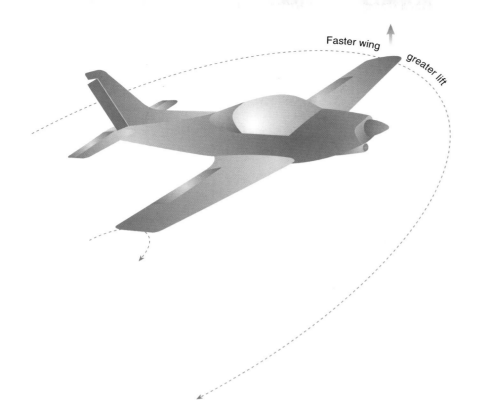

Faster wing

greater lift

Figure 79. When yawing, the outer wing moves faster than the inner wing, giving greater lift on that side and, as a secondary effect, causing the aircraft to roll, LEFT.

Figure 80. The action of the elevator trim tab in reducing the load on the stick, BELOW, *allows the pilot to fly with "hands off." The airflow over the surfaces of the tab holds the elevator in any given position.*

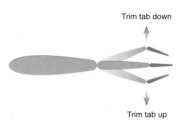

Trim tab down

Trim tab up

Once again, a spiral dive will develop as the nose follows the lower wing downward. So although rolling and yawing have their separate controls, the two motions are inseparable. To execute a normal balanced turn, the pilot must apply aileron and rudder together in the same direction.

TABS & TRIMMING

Any change in the weight distribution, whether because of passengers, baggage or fuel, alters the balance of the aircraft. Changes in throttle and airspeed also upset the balance, and the pilot must correct the unbalanced forces that result by adjusting the controls. In normal circumstances, no great physical exertion is required to operate the stick or the rudder bar, but on a long journey or during an extended climb, it would become extremely tiring to have to hold the control in the necessary position, especially if conditions were turbulent. The purpose of trimming is to reduce the load on

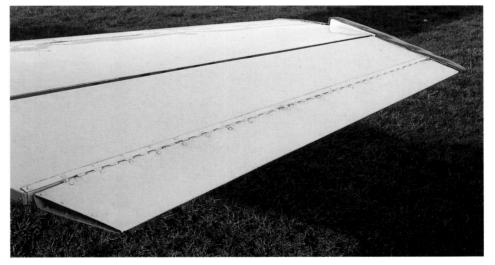

the controls, allowing the pilot to fly with "hands off" in any attitude. There are a number of methods of achieving this, but the most common is to have a small adjustable surface, known as a trim tab, on the control surface in question (see photo above and Figure 80).

Most aircraft are fitted with an elevator trimmer, but generally, only the multiengine types have rudder and aileron trimmers. The elevator trim tab is mounted on the

The elevator showing the trim tab—the hinged surface at the trailing edge.

The Airbus 340—a state-of-the-art airliner of the late 20th century—has advanced aerodynamics and fly-by-wire control systems.

elevator's trailing edge and can be adjusted by a control wheel in the cockpit in such a way that the airflow over the surfaces of the tab holds the elevator in any position without the slightest effort on the part of the pilot.

STRAIGHT & LEVEL FLIGHT & SPEED RANGE

In straight and level flight, there is an equilibrium of forces whereby weight is balanced by lift, and thrust is balanced by drag. A change in any one of these leads to a change in the others. If power is increased by opening up the throttle, for instance, the thrust will become greater than the drag, so the aircraft will accelerate. The resulting increase in speed will generate more lift, and the aircraft will climb. The speed of an airplane, unlike that of a vehicle on the ground, is controlled primarily by the elevators, not the throttle, so to increase airspeed, the angle of attack is reduced by lowering the nose in relation to the horizon. The chief purpose of the throttle is to control the rate of climb or descent at any particular airspeed,

but the throttle and elevators must be combined in such a way as to achieve the desired effect, so if the nose is lowered to increase speed, more throttle may be required to maintain height. If, on the other hand, the speed is reduced by pulling the stick back, the machine will usually climb, unless the engine power is decreased.

One of the most important qualities of an airplane is the range of speeds within which it can remain in straight and level flight. Some of the earliest aircraft had little speed range at all, owing to their limited engine power. If they flew at anything less than full power, they lost altitude. But nowadays, many aircraft have a maximum speed five times greater than their minimum. To fly at maximum speed and still maintain level flight, the pilot must keep the lift equal to the weight. As the form drag and skin friction resulting from the faster airflow build up, more engine power is required. Eventually, a point is reached at full power when maximum speed in level flight has been attained; any further increase in airspeed can be gained only by losing height.

To fly as slowly as possible, the angle of attack has to be increased to maintain height. The minimum speed for level flight is reached when the induced drag becomes so high that even at maximum throttle, the aircraft is unable to maintain altitude, although the stalling angle could, of course, be reached before this point. It is interesting to note that aircraft, birds and insects differ from a vehicle moving on the ground in that they must all work at high power in order to move as slowly as possible.

Between the two limits of the

speed range, there is a speed at which an aircraft in level flight flies with the least possible thrust and minimum drag; this is at the best lift-to-drag ratio, when the aircraft has attained the most efficient cruising speed. The most efficient speed for jets, however, can be up to 30 percent faster than the speed for minimum drag.

FLAPS, SLOTS, STALLING & SPINNING

An aircraft's speed range can be considerably extended at its lower end by devices such as flaps and slats. Flaps were developed before the First World War and today come in various forms (see Figure 81). They are usually positioned in-board of the ailerons, from which they can be extended or withdrawn

into the wings' trailing edges. They work by increasing the camber of the wings, and in the case of Fowler flaps, there is the benefit of the extra area as well as the slot effect. With some aircraft—the Boeing 747, for example—the extent of the flap movement is considerable, increasing the total area of the wing by about one-quarter, as can be seen when this huge aircraft makes its landing approach (see photo above).

Flaps are an enormous advantage when landing, as they increase both lift and drag. By boosting lift, they significantly reduce the stalling speed, thereby permitting lower landing speeds. By increasing the drag, they help in two ways. First, they enable the aircraft to descend steeply at the required approach speed, allowing for better obstacle

Fowler flaps fully extended on the wing of a Boeing 747 show the three sections and triple slots. This device reduces the stalling speed from about 240 knots to 140 knots.

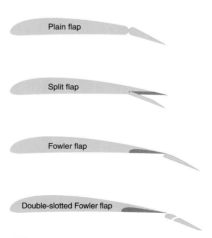

Figure 81.
Several types of flaps.

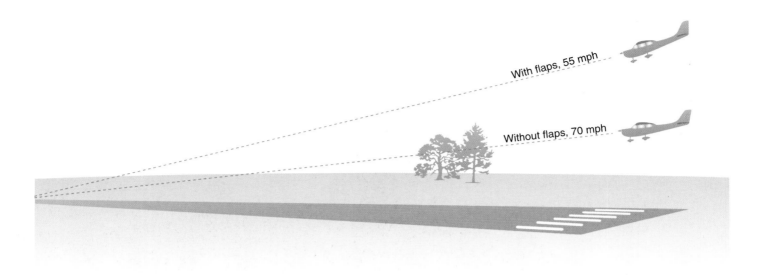

With flaps, 55 mph

Without flaps, 70 mph

Figure 82A. Flaps play a vital role in the landing of an aircraft: They enable the pilot to descend steeply without increasing speed, providing improved obstacle clearance, better forward visibility and lower stalling speed.

Figure 82B. The benefit of flaps and slats on lift and stalling angle is depicted in this graph.

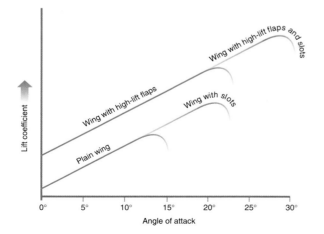

clearance. Second, the flap extension usually makes the aircraft adopt a more nose-down attitude, providing the pilot with better forward visibility (see Figures 82A and 82B).

An alternative method of steepening the approach path is by artificially increasing the aircraft's drag through the use of air brakes, or spoilers. There are various types, and they can appear as flaps, sometimes on the fuselage but more often above or below the wings. Air brakes differ from slots in that they neither increase lift nor reduce stalling speed and so have no effect on an aircraft's speed range, although their exact function may vary according to the type of air-

craft. They may be used to check speed before turning, to assist lateral or longitudinal control or to kill lift and slow the aircraft after landing. They are particularly important on sailplanes and other "clean" aircraft. By spoiling the lift-to-drag ratio and steepening the glide angle, the aircraft descends rapidly and is prevented from floating on.

For millions of years, birds have been using their slotted wings for slow-speed flying; the British firm of Handley Page Ltd. was the first to adopt this principle when it developed slotted wings for its aircraft in 1919. The actual slot is a gradually narrowing gap on the wing's leading edge and is given its shape by a small auxiliary airfoil known as a slat. Analogous to the bird's alula (see Figure 60A), the slot functions by directing a vigorously moving stream of air over the wing's upper surface to reinforce the boundary layer; by so doing, it smooths out turbulence and postpones the stall. It can increase the stalling angle to 25 or even 30 degrees and can sometimes increase the lift by as much as 100 percent.

In light aircraft, the slats usually work automatically, lying flush along the wing's leading edge at high speeds and pulled forward at low speeds as the angle of attack increases. But in larger aircraft, the slats are more often linked mechanically with the lowering of the flaps. As well as reducing the stalling speed for a given wing area, flaps and slots provide extra lift, which has permitted the use of smaller and therefore lighter wings.

Birds, bats and insects do not suddenly fall out of the sky as a result of stalling or spinning. In addition to making use of nonsteady airflow and variable-geometry flying surfaces (birds and bats), animals are equipped with a highly evolved nervous system that allows instinctive monitoring of flight conditions. With aircraft, however, stalling has always been one of the greatest perils of flying. The danger lies in the loss of height that follows the stall, because it is impossible to recover control until sufficient airspeed has been regained, which may take several hundred feet. Stalling occurs at high angles of attack, when the airflow breaks away from the wings' upper surface. The most dangerous and, unfortunately, the most frequent stalling occurs shortly after takeoff or before landing, when airspeed is low and there is insufficient altitude in which to recover. The development of flaps and slats, however, has led to an enormous improvement in air safety for both large and small aircraft, not only permitting lower landing speeds but also sometimes eliminating the conventional stall altogether.

In the typical stall, the center of pressure moves suddenly backward, upsetting the aircraft's balance (see Figure 83A). As a result, the nose drops forward, the tail pitches up

and the aircraft rapidly loses height, in spite of the stick being held fully back. In light aircraft, the pilot is warned of an approaching stall by floppy controls, or buffeting, so corrective action can be taken. Control can be effected simply by easing the stick forward to regain flying speed, applying power and then bringing the nose back up to the horizon again. In modern jets, sophisticated stall-recognition and prevention devices have almost eliminated the problem.

At any given angle of attack, an aircraft that is heavily loaded must fly faster than an empty one of the same type, because the wing-loading is greater. Consequently, the stalling angle will be reached at a higher speed. The stalling speed also becomes higher when the effects of g (force of gravity) temporarily increase the wing-loading, as in steep turns and aerobatic maneuvers, resulting in a sudden and often alarming high-speed stall.

A spin, or autorotation as it is sometimes called, is an unstable stall in which the aircraft is simul-

The leading-edge slats open, revealing the slot. Leading-edge slats usually work automatically, being "sucked" forward at low speeds as the angle of attack increases.

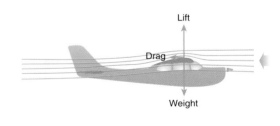

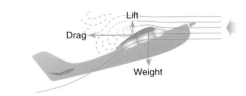

Figure 83A. Airflow during normal flight and the stall showing the change in lift, drag and center of pressure.

*Figure 83*B. *An aircraft in a spin.*

opposite aileron. This makes the condition worse, however, because it tends to increase the angle of attack of the stalled wing, adding to drag and thus causing more yaw. The only way to recover from a spin is to stop the aircraft from yawing by applying full rudder in the opposite direction to the spin, then to eliminate the stalled condition by easing the stick forward—an action a pilot should practice to gain a suitable level of self-confidence.

For years before it was understood to be one of the possible consequences of stalling, the spin mystified pilots and caused an enormous number of flying accidents. In the early days, aircraft would enter a spin, apparently for no reason, and there seemed little the pilot could do to recover. Now that we understand its origin and how to effect recovery, the spin is far less dangerous.

THE SOUND BARRIER

Toward the end of the Second World War, American pilots flying P-38 Lightnings and P47 Thunderbolts and English pilots flying Spitfires and Typhoons experienced frightening buffeting when they put their aircraft into steep powered dives. The vibrations were sometimes so violent that the wings disintegrated, tail units broke away from the main structure and pilots lost their lives. What they did not know then was that their aircraft were flying so fast, the air was unable to move out of the way quickly enough and became compressed, forming shock waves which hammered on the aircraft's structure.

Until that time, aircraft had been designed to function only at speeds of up to about 450 miles per hour,

taneously pitching, yawing and rolling. It is initiated by the onset of a stall, when one wing loses lift before the other, perhaps caused by entering the stall with an unintentional yawing motion. As the wing drops, the angle of attack increases and aggravates its stall. At the same time, the ascending wing becomes partly unstalled because of a greater relative airflow from above; the roll and resulting sideslip swing the aircraft into a yaw. If left uncorrected, the nose will drop and follow the fully stalled wing, and a spin will develop and continue until the pilot takes suitable action.

The pilot's natural reaction is to raise the lower wing by applying

and little was known of the aerodynamic forces that occurred at higher speeds. As the speed of sound is approached, the behavior of air changes drastically and several of its aerodynamic characteristics work in reverse. A solution to the problem became particularly urgent, as it presented a barrier to the higher-speed flight made possible by the new jet engines. Of course, there was no physical "sound barrier," as such; it was more in the minds of those who thought that the speed of sound was the limiting factor determining man's future progress in the air. Nevertheless, a number of pilots—notably Geoffrey de Havilland, flying the tailless DH108 Swallow—perished in the attempt to "break" the sound barrier.

Until comparatively recently, it was extremely difficult to study high-speed airflow in the laboratory, because wind tunnels became choked at supersonic speeds. The only way to investigate supersonic airflow was to risk actually flying through the barrier. To this end, the U.S. Air Force ordered a special rocket-powered research aircraft, the Bell X-1, designed to withstand the severe buffeting associated with speeds of around the speed of sound. In October 1947, it was air-launched at 30,000 feet beneath the fuselage of a B-29 Superfortress. After a series of test flights, during which speeds were gradually increased, pilot Chuck Yeager, in spite of suffering severely from a horse-riding accident the previous day, finally applied full power. When he approached the critical speed, the machine began to shake alarmingly, nearly going out of control. Then, quite unexpectedly, the buffeting stopped—Yeager had passed through

the barrier into supersonic flight.

So what are the characteristics of supersonic airflow that make it so different from ordinary airflow? Although the subject of supersonic aerodynamics is highly technical when studied in detail, a book on the evolution of flight would be incomplete without at least some explanation, and this simple analogy may help. When a car is driven slowly through a flock of sheep, the animals saunter to each side in their own time and in an orderly fashion. However, if the car is driven faster, the animals, rather than being persuaded to move out of the way more quickly, bunch up and are unable to adjust their flow pattern to suit the speed of the car. Similarly, when an object is moving through the air at around the speed of sound, the air has insufficient time to adjust itself to the new flow. Its molecules are squeezed together and increase in density before they move aside to allow the object to pass.

When an object moves through the air at less than the speed of sound, the air in front of it is warned of the impending disturbance by pressure waves. Whether audible or not, these waves are the same as sound waves and are a succession of weak increases and decreases in the density of the air. Traveling ahead of the flying object, the pressure waves warn the air that, for instance, there is lower pressure above the wing than below and that it is easier to pass over the

*Figure 84*A. *A conventional subsonic aircraft wing moving at about Mach 0.5 (less than the speed of sound) shows the airflow breaking away.*

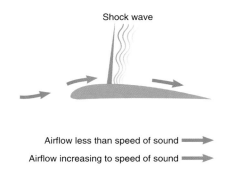

Shock wave

Airflow less than speed of sound ➡
Airflow increasing to speed of sound ➡

Figure 84B. At about Mach 0.75 and increasing to the speed of sound, we see the incipient shock wave.

top. As a result, the air begins to move aside some distance before the wing meets it, most of it curving over the wing's top surface. The warning effect can easily be seen in many of the airflow diagrams. Pressure waves travel at the same speed as sound; so if the aircraft is also flying at the speed of sound, the warning is clearly unable to arrive before the air strikes the machine. This warning is crucial, because it influences the behavior of the air when it meets the aircraft; the difference in speed between the aircraft and sound is therefore a key factor.

TRANSONIC & SUPERSONIC FLIGHT

The speed of sound varies according to the temperature of the air: In average temperatures at ground level, it is around 750 miles per hour, but in cold conditions above 36,000 feet, it is about 660 miles per hour. The higher the temperature, the faster sound travels. From an aerodynamic point of view, then, the actual speed of the aircraft is not as important as its speed relative to the speed of sound in the ambient conditions in which it is traveling. This relationship is expressed in Mach numbers. An aircraft flying at Mach 0.5 is moving at half the speed of sound (anywhere from 330 to 380 miles per hour). It is sometimes more convenient to speak in the broader terms of subsonic, sonic and supersonic speeds, rather than specifying the actual Mach number. When an airplane moves from subsonic to supersonic speeds, the transition is generally gradual rather than instantaneous. The range of speeds within which this occurs is known as the transonic region. The slow side of the

barrier has the more alarming effects on aircraft and, furthermore, has proved the most difficult obstacle for designers. Even now, less is understood about flight at transonic than at supersonic speeds.

As speed increases, the first sign of change in the nature of flow is the breakaway of the smooth airflow over the surface of the wing or other part of the aircraft. This may occur when the aircraft is moving at about half the speed of sound, but as the speed increases, the point of breakaway creeps forward, resulting in a thicker turbulent wake and a severe increase in drag (see Figure 84A).

A stage is reached at about three-quarters of the speed of sound when a new phenomenon occurs: the incipient shock wave (see Figure 84B). It usually arises at the point of maximum camber, where the air passing over the wing reaches its greatest speed and appears as a line at right angles to the surface, marking the position where the air pressure, density and temperature have reached a critical level. Here, the local airflow is already at Mach 1, and the speed at which this first shock wave forms is called the critical Mach number. The less slender the aircraft, the lower the critical Mach number.

Smaller shock waves may also form on other parts of the machine —at the entrance of the jet engine, for instance, where the local airspeed reaches the speed of sound. One of the aims of the aircraft designer is to minimize local shock waves and to keep the critical Mach number high so that the transition from subsonic to supersonic flight is as smooth and rapid as possible.

The aerodynamic forces generated by shock waves are very strong and lead to an enormous increase in drag and a reduction in lift. Whereas most of the energy in low-speed flight is consumed in overcoming induced drag, most of the energy at high speeds is lost in the form of heat dissipated by shock waves. The airflow behind the shock waves becomes extremely turbulent, because the boundary layer separates from the wings' surface, resulting in a loss of lift and a further huge increase in drag. The shock wave and the resulting turbulence produce shock, or wave, drag, increasing the resistance of a conventional airfoil by many times. To offset this, a corresponding increase in engine power is needed.

The graph below shows the large increase in drag during the transition from subsonic to supersonic flight, but once the transonic region is left behind, the relative stability of supersonic flight takes over and drag diminishes significantly. The actual speeds that separate these regions are difficult to specify, since so many factors come into play. Perhaps the best definition is to say that in subsonic flight, airflow over all parts of the aircraft is subsonic; in transonic flight, some of the airflow is subsonic and some supersonic; while in supersonic flight, all the airflow is supersonic.

At one time, the prospects of supersonic flight seemed daunting. Apart from its physical effect on the aircraft, there was a lack of suitably powerful engines to overcome the drag increase at transonic speeds. The effects of speed, drag and compressibility of air can be gauged by the fact that about 30 times more power is needed to fly at 750 miles per hour than at 300 miles per hour. Moreover, the issue is exacerbated with propeller-driven aircraft, chiefly because the combination of the rotary speed of the propeller and the high forward speed of the aircraft is sufficient to cause the blade tips, where the speed is highest, to suffer the effects of compressibility and shock waves. This starts to occur at speeds of about 400 miles per hour, but it rapidly builds up, becoming so catastrophic that at 750 miles per hour, the power needed to sustain flight in a propeller-driven aircraft is some 120 times that required at 300 miles per hour.

Fortunately, Sir Frank Whittle came to the rescue by inventing the jet engine. At a stroke, this new source of power eliminated the problems associated with the propeller. In addition, its power-to-weight ratio was a dramatic improvement on the piston-propeller engine, particularly as the efficiency of the gas turbine increases rapidly when that of the propeller is quickly falling. Thus the jet engine marked the first step in solving the seemingly intractable dilemma facing supersonic flight.

As the speed of the wing increases beyond the critical Mach number, a second shock wave

TRANSONIC DRAG & THE CONCORDE

There is an enormous increase in drag generated at transonic speeds, as illustrated in the graph BELOW. *To a remarkable degree, this is demonstrated by the Concorde as afterburner, or reheat, which increases engine power by 20 percent and is required to overcome the drag increase in the transonic region. At speeds in excess of Mach 1.7, however, the aircraft accelerates with reheat switched off (in dry thrust). In this respect, the Concorde remains unique among all other supersonic aircraft.*

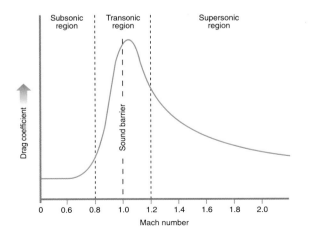

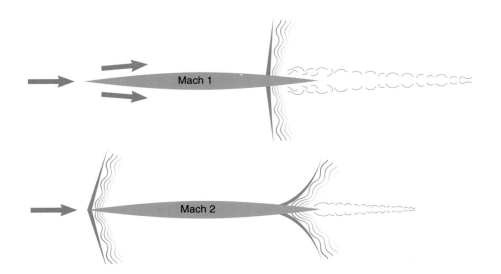

Figure 86A. The change in shock waves at speeds increasing from Mach 1 to Mach 2.

TUCK-UNDER

In the early days of transonic flight, one of the main dangers of flying at close to the speed of sound was "tuck-under." As a result of the center of pressure moving backward, the nose dropped, and when the pilot tried to oppose this by pulling back on the elevators, the problem was only exacerbated. The Bell X-1 (see page 155) utilized an all-moving tailplane rather than conventional elevators, and this greater power enabled better control to be maintained beyond Mach 1.

Figure 86B. Airfoils suitable for supersonic flight.

forms on the lower surface, and both shock waves move downstream as the airspeed continues to increase (see Figure 86A). At Mach 1, the waves attach themselves close to the wing's trailing edge, and when the speed of the aircraft exceeds Mach 1, an additional wave, the bow wave, forms in front of the leading edge. Thereafter, apart from the angles becoming more acute with increasing speed, there is little change in the shock wave pattern.

It may come as a surprise to learn that once the transonic region has been passed, the conventional aerodynamics of low-speed flight bear little resemblance to the aerodynamics of supersonic flight. The boundary layer is now relatively unimportant, as the viscous forces within it are small—this largely accounts for the capacity of supersonic flow to turn sharp corners. Curiously enough, at much higher speeds, the boundary layer assumes more importance as it thickens once again. Another strange aspect of supersonic flight is that the shape of the airfoil is not critical as long as it is thin. Straight lines and sharp corners are as good as or even better than curved surfaces. Symmetrical bicon-

cave, wedge-shaped or hexagonal airfoils are all excellent for supersonic flight.

Slimness, then, is one of the chief characteristics of aircraft with high critical Mach numbers. An interesting example of this was the Spitfire, a slim aircraft of its time that was designed with little thought to flying close to the speed of sound. Yet it was eventually shown to have one of the highest critical Mach numbers achieved—nearly 0.9.

SUPERSONIC AIRCRAFT, VORTEX GENERATORS & VARIABLE GEOMETRY

How did the physics of supersonic flow affect the evolution and design of high-speed aircraft? Early designers faced a formidable task. They had learned that the flying characteristics of an aircraft designed for subsonic flight changed dramatically—even dangerously—as it entered the transonic region. The machine was thrown out of balance by the center of pressure moving rearward at the onset of the shock wave, which had an alarming tendency to oscillate to and fro on the wings, interfering with the boundary layer and the airflow on the tailplane. The condition, known as a shock stall, can occur to a lesser or greater degree before the aircraft as a whole has reached the speed of sound.

As a result, the machine was subject to violent vibrations and buffeting. Even worse were the effects of shock waves forming on the tailplane, which led to the elevator controls becoming so stiff that they effectively became jammed and, on occasion, reversed the controls of the stick. Unlike the conventional slow-speed stall, however, the shock stall may occur at any angle

of attack, although it is most likely to occur at the small angles associated with high speeds. Recovery is made by slowing the aircraft down using the air brakes, closing the throttle and, if the elevators can be moved, pulling the nose up.

Designers appreciated the importance of thin, streamlined shapes for minimizing drag at speeds of around 400 miles per hour, which were then considered high. As engineers began to understand more about shock drag and the origin of shock waves, it gradually became clear that airfoils of symmetrical design with low ratios of thickness to chord were essential and that bulges, bumps and cambered surfaces should be avoided. Complications arose in that the design of aircraft intended for supersonic flight could hardly be less suited to the low speeds essential for takeoff and landing, and in time, the design of supersonic aircraft assumed an entirely different form from that of subsonic aircraft. In contrast to subsonic aircraft, which have smooth, rounded contours, supersonic aircraft became angular and dart-shaped (see Figure 86B and photo on page 161).

Although the Spitfire was designed in the 1930s, it was eventually shown to have achieved the highest critical Mach number for its time.

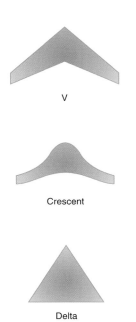

V

Crescent

Delta

Figure 87. Swept-back wings delay the onset of the critical Mach number, allowing jetliners to fly at speeds approaching the speed of sound.

Figure 88. Nipping in the waistline can improve an aircraft's performance at high speeds.

One of the most significant contributions to high-speed flight was the application of sweepback (see Figure 87). Slightly swept-back wings had already been used in aircraft to add stability and improve the pilot's view from the cockpit, but by substantially increasing the angle of sweepback to between 40 and 70 degrees or more, the critical Mach number of the wing can be raised, delaying the onset of shock stall and sometimes avoiding it altogether.

The wing's geometry is so arranged that a lower air velocity is obtained at right angles to the leading edge (across the chord of the wing), for this is the velocity which largely determines the wing's critical Mach number. Aircraft with swept-back wings can be made to fly at much higher speeds without producing shock stall than those with relatively straight wings, allowing, for instance, today's jetliners to fly at speeds of around Mach 0.9. Performance at these speeds can also be improved by nipping in the waistline to avoid the bulges that cause shock wave and drag according to the so-called area rule (see Figure 88). This principle means that the cross section of the aircraft is kept more constant from nose to tail, offsetting the additional area presented by the wings.

One of the effects of transonic flight is that the boundary layer becomes sluggish over the rear part of the wings, causing it to thicken and break away from the wings' surface. Various methods are used to control this separation and act by giving the boundary layer a new lease on life, often reducing the intensity of shock waves and wave drag. Vortex generators, which consist of projections on the upper surface of the wings

(see photo on page 163), provide the simplest way to reinforce the boundary layer by increasing the turbulence and energy of the boundary layer. In some respects, vortex generators have an effect similar to that

*The modern supersonic fighter
Saab Gripen.*

The Super VC 10 is one of the most beautiful subsonic passenger aircraft ever designed. The weight of its four engines is at the rear, and its swept-back wings are placed much farther back than they are in aircraft with wing-mounted engines. Designed for short-field performance at high-altitude airfields, rather than for economy, this superbly made aircraft has been out of commercial service since the 1980s but is still used by the Royal Air Force.

of leading-edge slots. Other devices that achieve the same end include boundary-layer fences on the wing, sawteeth on the wing's leading edge and various more complex blowing and suction arrangements.

The versatility and speed range of aircraft intended for flight at high Mach numbers can be improved by imitating the high-speed flight of diving birds, such as falcons and gannets, which fold their wings toward their bodies with the tips swept back during a dive and extend them outward during takeoff and landing. The idea of variable-geometry wings in aircraft was originally proposed in the 1940s by an Englishman, Dr. Barnes Wallis. His basic notion was simple: At high speeds, the wings assume a delta shape, while at low speeds, to provide the

extra lift and a sufficiently low induced drag, the wings pivot outward, increasing both the area and the aspect ratio (see Figure 89).

This system entails a host of practical and mechanical challenges, such as the strength required by the pivots that must bear enormous aerodynamic loads and the complications of redistributing the weight to maintain trim as the position of the wings is adjusted. Nevertheless, this system has been successfully employed by a number of military aircraft, including the F1-11, the B1 bomber and the TU 160 Blackjack.

DELTA WINGS & THE CONCORDE
The most popular plan form for supersonic flight is the delta shape. It has high-speed qualities similar to

Vortex generators on the wing of a Raven microlight. These projections act in a way similar to leading-edge slots by livening up the airflow in the boundary layer and preventing separation.

those of swept-back wings but is stronger and provides longer trailing edges on which control surfaces can be mounted. But to produce the necessary lift at low speeds, delta wings also require a higher angle of attack than conventional aircraft, as can be seen in some fighter aircraft and in the Concorde when landing and taking off.

Fortunately, an unexpected bonus arises with the delta wing at high angles of attack that does not occur with conventional wings. It bears a striking similarity to the "party-streamer" effect recently discovered in the flight of insects (see page 72). Some of the air at the sharp leading edge separates to form a concentrated vortex that flows backward over the wings' surface; at the same time, the boundary layer at the rear tends to flow outward. By modifying the plan form of the delta, this leading-edge vortex can be persuaded to adhere to each wing without breaking up, even at slow speeds up to and beyond the stall. This can be enhanced by mak-

ing the wings' leading edges curved with rounded tips and by increasing the sweepback from extended roots.

Such a form gives the Concorde its striking Gothic appearance and can improve an aircraft's lift by over one-third. In fact, the Concorde is so designed that it does not need any extra lift-producing devices such as flaps or slats.

In common with many delta-wing aircraft, the Concorde does not have a tailplane. Although the tail plays an essential role in the maintenance of stability in conventional aircraft, there are several good reasons for doing away with it in high-speed machines. By eliminating as much as possible of that which does not contribute to lift, parasitic drag can be kept to the minimum. Without a tail or elevators, the aircraft becomes a "flying wing," which means that some alternative way must be found to maintain longitudinal stability and control pitch. As any control-restoring surface needs to be well behind the center of

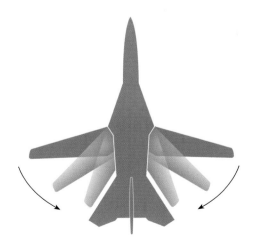

Figure 89. The principle of the variable-geometry wing was first advanced by Dr. Barnes Wallis during the 1940s.

The striking Gothic shape of the Concorde was the result of 4,000 hours of wind-tunnel testing and has produced an aircraft so aerodynamically efficient that high-lift devices such as slats and flaps are not needed. Much of the space-shuttle technology stemmed from the Concorde.

gravity, the obvious place for it is on the tips of the swept-back delta wings, so the wingtips act in the same way as a tailplane (see page 27). The trailing edges of the wings are also provided with elevons, which take the place of the elevators. These combine the functions of the elevators and ailerons of conventional aircraft.

The Concorde's control and stability in the transonic region is improved by having an adjustable center of gravity—perhaps one of the few characteristics it shares with

the hang glider. As the center of pressure moves back at transonic speeds, fuel is pumped from trimming tanks at the front of the fuselage to a trim tank at the rear of the machine, where it remains during supersonic cruise. When the Concorde slows down to subsonic speeds, the center of gravity is brought forward again by fuel being returned to the forward fuselage tanks. This fuel is also used as a heat sink to absorb the high skin temperatures generated at high speed. The brilliant design of the

Concorde's fuel system, with its 13 tanks, is considered by many to be the heart of the aircraft.

Perhaps the greatest virtue of this aircraft is its basic simplicity. By limiting the Concorde's speed to around Mach 2, the aircraft's skin temperature does not rise to extreme levels, permitting aluminum construction rather than the heavier, more costly and hard-to-handle titanium. The Concorde also has the advantages of both deep chord and long span. The deep chord provides sufficient strength for thin wings, which are so vital in transonic and supersonic flight, while the long span keeps induced drag to a minimum. Not least among this aircraft's qualities is its seductive yet restrained three-dimensional ogee shape that allows high angles of attack in almost perfect safety. These attributes, together with the complete lack of such superfluous attachments as spoilers, flaps, slats, tailplane and elevators, all help to make for a truly remarkable and beautiful aircraft, which for over 30 years has been thrilling everybody who sets eyes on it, let alone those who are fortunate enough to fly in it. In the United States, the Concorde is considered second only to the space shuttle, which, by the way, was designed using Concorde know-how, procedures and aerodynamics.

HIGH MACH NUMBERS & HYPERSONIC FLIGHT

The laws of aerodynamics are perverse. No sooner do we solve the problems of subsonic flight than we enter into another region of speeds, where we face the fresh challenges of transonic and supersonic flight. After solving most of these, we find that at Mach 5, the rules of "conventional" supersonic flight are transformed, and we pass into the strange world of hypersonic flight.

To begin with, the boundary layer becomes thicker once again, while the angle of the shock wave becomes more acute. The shape of the airfoil, on the other hand, assumes even less importance than it did in supersonic flight. And then there is the vexing question of kinetic heating from skin friction, which, at high Mach numbers, can raise the temperature to 9,000 to 18,000 degrees F. Even more fundamental is that at around Mach 9, the very nature of air changes, and its molecules of oxygen and nitrogen dissociate into atoms and ions. This phenomenon is responsible for the disruption of radio communications when spacecraft reenter the Earth's atmosphere.

The difficulties of hypersonic flight include finding suitable materials that retain adequate strength at high temperatures, preventing the fuel from boiling in the tank and insulating the passengers and crew from the searing heat. There is also the matter of developing suitably powerful and efficient engines. A new type of jet, the scramjet, may provide the answer. Based on the simple ramjet, it is designed to function at hypersonic speeds, burning hydrogen.

The significance of this new engine is twofold. A conventional rocket engine must carry a supply of oxygen to burn its fuel, but the scramjet scavenges oxygen from the air as it flies, thus saving considerable weight. The higher it flies, the thinner the air, but this is naturally compensated for by the higher hypersonic speed and corresponding increase in oxygen flow compressed through the engine. At the same

Figure 90A.
The shape of things to come.

time, the rarefied atmosphere also helps to reduce the intense friction-generated heat. The shape of such futuristic aircraft is uncertain, as comparatively little is known about hypersonic aerodynamics. Although a wedge shape is one possibility, consideration is now being given to a wingless lifting body—perhaps pebble- or flying-saucer-shaped.

The faster we fly and the more we learn about aeronautical science, the less "real" our airplanes seem to become. We need only look at the atrophied wings of some supersonic aircraft to realize that with increasing speed, our flying machines begin to function more like rockets, relying on engine thrust rather than the lifting properties of wings. Although flying is now a safer activity than ever before, there is no doubt that like so many other aspects of our lives, it is also becoming more impersonal and remote for both pilots and passengers. Increasingly, pilots are becoming systems managers instead of airmen in the true sense of the word, relying on a profusion of navigational and computerized equipment rather than concentrating on the air, clouds and sky around them.

For passengers, too, the sensation of flight is fading. Crammed together like sardines in a giant cylindrical tin, we are encouraged to eat, drink, listen to piped music and watch second-rate films. Perhaps dullness is the price we have to pay for safety.

HANG GLIDERS & MICROLIGHTS

It is hardly surprising that in these circumstances, hang gliding and microlighting, or ultralighting, have now become such popular pastimes. We seem to have come full circle, reverting to leaping off hillsides and cliffs on fabric wings like the medieval tower jumpers.

Hang gliding is the simplest and cheapest form of flying and is available to anyone who is reasonably fit and bold enough to try. Hang gliders can be flown from any suitably sloped hill facing the prevailing wind. The sport is quiet and uncomplicated. It has few rules or regulations, does not need runways,

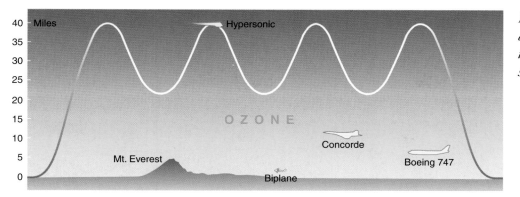

Hypersonic

O Z O N E

Concorde

Mt. Everest

Boeing 747

Biplane

*Figure 90*B. *In the future, hypersonic aircraft may surf through the air on their own shock waves in a series of skips 40 miles high.*

radios, control towers or winches and leaves behind no traces. Above all, it is exhilarating and gives one a greater feeling of being in harmony with nature than does perhaps any other form of flying.

One reason for this is that the pilot is an essential part of the structure, weighing about four times as much as his "wings," which become virtually an extension of himself. Unlike with other flying machines, the pilot has a bird's-eye view of the ground beneath as he wheels and soars almost with the grace and freedom of a gull. Perhaps, too, like mountaineering, the element of risk adds to hang gliding's appeal. Although a hang glider is, in some respects, easier to fly than light aircraft, it does demand a degree of physical fitness, and in the early stages of learning, the sport requires considerable courage and determination, as can be testified to by the author.

Experienced pilots with advanced hang gliders and parasails can today achieve spectacular feats that were once thought possible only with powered aircraft and sailplanes. Given the right conditions, pilots can remain airborne for several hours, ascend in thermals to thousands of feet and make long cross-country journeys. Some can

even perform certain aerobatic maneuvers, although such stunts carry grave risks. The remarkable feature is that all this is accomplished with an aircraft which can be folded up, carried on the shoulder and stored in the garden shed.

Hang gliding, of course, started with Otto Lilienthal in the 19th century, the point at which successful manned aviation began. The difference between his attempts and those of his present-day followers is that Lilienthal was not only testing and developing his glider but also learning to fly without the benefit of the previous experience of others. Most modern flex-wing hang gliders date back to 1951, when American Dr. Francis Rogallo patented a delta-wing kite with flexible surfaces whose shape was maintained by tension wires.

As part of a multimillion-dollar research project for the recovery of space capsules, the U.S. Space Agency developed and thoroughly tested the delta-wing kite in wind tunnels. It was shown to possess, among other characteristics, a very high level of stability. When the project was subsequently abandoned, several people were quick to realize the kite's recreational possibilities. In the late 1960s, Americans Michael Markowski,

RAMJETS

The original ramjet was a simple jet engine that consisted of a specially shaped tube within which fuel was burned. The forward motion of the aircraft collected and compressed the air, which was then heated by the burning fuel, resulting in the hot gas being shot out at high speed. There were no compressor blades or other moving parts, and once the aircraft was under way, enough compression was generated to keep the fuel burning without any ignition device. However, there is one major snag —a ramjet functions well only when moving at speeds of at least 1,000 miles per hour. Thus an auxiliary means of propulsion is needed to get it up to speed.

A 1970s vintage Rogallo-type hang glider slope-soaring. Note the dart shape and the deep keel of the single-skin wing. Modern hang gliders have much higher-aspect ratios and double-skin wings and are thus more efficient.

an aeronautical engineer who worked on the design of the Douglas Aircraft Company's DC10, and Richard Millar, editor of *Soaring* magazine, saw a potential use for the Rogallo sailwing, as it was then called, for hang gliding. They adapted and developed it for foot-launched flight from the tops of sand dunes, and in due course, the design became refined and improved through the experience of others.

The modern hang glider has a flexible airfoil surface, known as the sail, made of fabric such as Dacron. This is supported by a tubular aluminum frame, which is strengthened by a number of tension lines. Because the Rogallo hang glider is so light, simple and stable, it requires no elevators,

ailerons or rudder and is controlled by the pilot's shifting his weight in relation to the sail. Although many hang gliders may not resemble conventional aircraft, they are subject to exactly the same laws of aerodynamics as any other subsonic flying machine. Directional stability is provided by the keel formed between the two cones of the sail, and the exceptionally high degree of lateral stability is attained by the weight of the pilot hanging underneath, giving pendulous stability. Sweepback on the wings normally ensures stability in the pitching plane, although reflex on the trailing edge of the inboard half of the wing is another way of achieving the same end (see Figure 23B and Figure 23c on page 27).

The pilot either sits in a seat or lies in a prone position, hanging in a harness attached to the top of a triangular-shaped control frame at the center of gravity. He controls the glider by holding the bottom of the frame and moving his body about. By pulling the frame toward him, the pilot shifts his weight forward in relation to the glider, so the nose drops and the airspeed is increased. If the pilot wishes to ascend or to reduce speed, he simply pushes on the frame. As these controls are opposite to those of an airplane, in which pushing forward on the stick causes the machine to pitch down, it can be confusing when a pilot of a conventional airplane takes up hang gliding for the first time—the consequences of pulling instead of pushing the bar during the final stages of landing need no further discussion!

Turns are also executed by shifting weight. If the pilot moves his body to the left, the shift in weight causes the glider to bank so that the lift acts sideways and the glider starts a left turn. In contrast to conventional aircraft, most hang gliders have mild stalling characteristics; if they are flown too slowly, they pitch down gently rather than suddenly going into a nosedive.

Although cross-country hang-glider flights are possible in places where strong thermals are plentiful, most flights in temperate climates are made by riding upcurrents at the tops of hills and cliffs, in the same way as do gulls and some

A basic "no frills" flex-wing microlight, the Medway Raven. Flex-wing microlights may look flimsy, but with their wire bracing, they are deceptively strong. They are also portable and have excellent short-field performance, as proved by the author after engine failure!

other soaring birds. In the absence of a suitable rising current, the pilot must content himself with a gradual descent to the bottom of the hill, from where he has to make the strenuous climb back, complete with his dismantled machine, before making another flight.

With glide ratios of only 5:1, the first hang gliders were very inefficient, but modern hang gliders have been much improved by increasing the aspect ratio and by using double-skinned airfoil-section wings. There is a limit, though, because if taken too far, extreme high-aspect ratios not only make the machines more difficult to handle, especially in high winds, but also require them to have control surfaces and other aids, which detract from the hang glider's principal attraction: simplicity.

Having to rely on rising currents of air to remain airborne can be frustrating when there are none, so it did not take long before more enterprising aviators attached a small engine to the hang glider to provide thrust, effectively turning it into a microlight. The first microlights were simply powered flex-wing hang gliders, but soon, new types emerged that were controlled via control surfaces and a stick rather than by weight shift.

Such three-axis machines look, function and fly more like conventional light aircraft and are generally easier to control in marginal weather conditions, particularly in crosswinds, but they are not without their disadvantages. They are much more complex, less tough (weight for weight) and less portable, require longer and smoother runways and are more expensive to produce. Clearly,

the chief advantage of the flex-wing type is the machine's simplicity and its ability to operate from any modest-sized rough field.

Although humans are, by our very nature, essentially earthbound, there is no longer any need for us to envy insects and birds in quite the way that our forebears did. With the artificial wings of aircraft, we can already outdistance and outpace all other living things. Our first wings, if they may be called such, were as unspecialized in their way as were those of the first flying insects 350 million years ago, but unlike them, our wings evolved in an infinitely shorter period of time—equivalent to a mere eight feet in a journey from London to Johannesburg. The fragile linen-covered surfaces on which the first forays into the air were made have developed within a few years into a bewildering number of forms, from the sleek metal airfoils that are the products of computers and advanced technology to the simple sails of hang gliders.

This book has sought to explain the basic principles that underlie all forms of winged flight, whether by insect, bird or man. The aerodynamics involved in lifting an object off the ground and keeping it in the air are universal, but one of the fascinations of the subject of flight is that there are still aspects of it which remain to be discovered and explored. Now that we have acquired experience with wings, we are better able to assess and understand the aerial performance of other fliers. Perhaps we can look at the wings of insects and birds with greater insight and appreciate the evolutionary processes that have made each creature perfect for its aerial role.

*A bird's-eye view of a fellow aviator.
To obtain this picture, the author
mounted the camera on the wingtip.*

Chanute, O. "Aerial Navigation." *Cassiers' Magazine*. 1901.

Clancy, L.J. *Aerodynamics*. Pitman, London, 1975.

Dalton, S.N. *Borne on the Wind*. Chatto and Windus, London, and Reader's Digest Press, New York, 1975.

D'Esterno, M.D. *Du Vol des Oiseaux*. Paris, 1864.

Ellington, C.P. and R.J. Wootton. "Biomechanics and the Origin of Insect Flight," from *Biomechanics and Evolution*. Cambridge University Press, Cambridge, 1991.

Ellington, C.P. et al. "Leading-Edge Vortices in Insect Flight." Letters to *Nature*, Volume 384, 1996.

Ennos, A.R. "A Comparative Study of the Flight Mechanism of Diptera." *Journal of Experimental Biology* 127, 1986.

Hecht, M.K. (ed.). "The Beginnings of Birds." Proceedings of the International Archaeopteryx Conference, 1985.

Hummel, D. *The Aerodynamic Characteristics of Slotted Wing-tips in Soaring Birds*. Verlag der Deutschen Ornithologen-Gesellschaft, 1980.

Kermode, A.C. *Mechanics of Flight*. 1972/1996.

Lilienthal, O. *Der Vogelflug als Grundlage der Fliegekunst*. R. Oldenbourg, Berlin, 1889.

Marey, E.J. *Le Vol des Oiseaux*. G. Mason, Paris, 1890.

Martin, L.D. *The Origins of Birds and Avian Flight*. Plenum Press, 1983.

Nachtigall, W. *Insects in Flight*. George Allen and Unwin, London, 1974.

Pedley, T.J. *Scale Effects in Animal Locomotion*. Academic Press, 1977.

Pennycuick, C.J. *Animal Flight*. Edward Arnold, London, 1972.

————. *Mechanics of Flight*. Avian Biology, 1975.

Pringle, J.W.S. *Insect Flight*. Cambridge University Press, Cambridge, 1957.

Rayner, J.M.V. "Avian Flight Energetics." *Annual Review of Physiology*, 1982.

Ward-Smith, A.J. *Biophysical Aerodynamics of the Natural Environment*. John Wiley & Sons, 1984.

Weis-Fogh, T. "Quick estimates of flight fitness in hovering animals." *Journal of Experimental Biology* 59, 1973.

Withers, P.C. *Wing-tips, Slots and Aerodynamics*. Ibis 123-239-247, 1981.

Wootton, R.J., J. Kukalova-Peck and D.J.S. Newman. "Smart Engineering in the Mid-Carboniferous: How Well Could Palaeozoic Dragonflies Fly?" *Science*, Volume 282, 1998.

Ader, Clément, 137

Aerial ship of 1670, 123; illustration, 122

Aerial Steam Carriage, 128-129; illustration, 129

Aerodynamics, principles of, 14-35

Aeshna cyanea, photo, 37

Ailerons, 147, 148; illustrations, 148

Air, 18-19
    compressibility, 18, 22
    density, 18, 19
    movement, 18-19
    pressure, 18-19, 23; illustrations, 22, 23
    viscosity, 28

Air brakes, 152

Airbus, 340; photo, 150

Aircraft. See also Balloons; Gliders; Hang gliders; Microlights; Ornithopters
    19th century, 125-143
    20th century, 144-171
    balloonlike inventions, 123
    earliest times, 116-118
    early fixed-wing, 128-130, 131-143
    engines, 146, 157, 165, 167
    history, 116-143
    hypersonic, 165-166; illustration, 166
    medieval times through 18th century, 118-125
    power-driven, development of, 137-138
    powering by human muscle, 121-122, 127
    supersonic, design considerations, 158-165

Aircraft control, 144-154
    control axes, 146; illustration, 146
    control surfaces, 147-149; illustration, 146
    flaps and slots, 151-153
    secondary interactions, 148-149
    trimming, 149-150

Airflow
    laminar, 29; illustration, 29
    nonsteady, 31-35
    nonsteady, and bird flight, 106-107
    nonsteady, and insect flight, 68-74
    supersonic, 155-156
    velocity, 19

Airfoils, 22-26; illustrations, 21, 23
    for supersonic speeds, illustrations, 158

Airplanes. See Aircraft

Albatross
    as inspiration for glider, 134; illustration, 135
    energetic cost of flight, 93
    gliding, 94, 98
    landing, 112-113

Alula, 90; illustration, 90

Angle of attack, 19, 20, 21, 23, 24; illustrations, 20, 23
    in insect flight, 58

Ara ararauna, photo, 13

Archaeopteryx, 12, 78-79; illustration, 78

Archimedes' principle, 16

Argentavis magnificens, 81

Aristotle, theory of flight, 14

Artingstall, F.D., 128

Aspect ratios, 25-26; illustrations, 25

Athene noctua, landing, photo, 114-115

Atmosphere, 18; illustration, 18

Autorotation, 153-154

B-29 Superfortress, 155

B1 bomber, 162

Bakker, Robert, 80

Balloons, 16; illustration, 16
    early balloonlike inventions, 123
    Montgolfier, 125; illustration, 124

Barrel roll of raven, illustration, 94

Bats, 32; photo, 32

Beech 2000A Starship 1, photo, 147

Bee-eater in flapping flight, photo, 100-101

Bees
    bumblebees, photos, 30, 60
    bumblebees, temperature control, 44
    coupling of wings, 50
    powering of flight, 51
    speed, 61
    wing-beat frequency and speed, 60

Beetles
    cockchafer, photos, 49, 66
    takeoff, 65-66
    use of single pair of wings, 50
    wing-beat frequency and speed, 60

Bell X-1, 155, 158

Bernoulli, Daniel, principle, 22-23; photo, 22

Bibliography, 172-173

Biplanes, photo, 14

Bird flight, 76-115. See also Bird wings; Birds; Feathers
    and nonsteady airflow, 106-107
    as model for manned aircraft, 125-126, 132-135, 139, 143
    clap-fling action, 107
    energetic cost, 92-93
    evolution of, 80-81
    fast, 102-103; illustration, 102
    flapping flight, 99-110; illustration, 102; photos, 100-101, 104-105
    fuel for, 88-89
    glide ratio, 94
    gliding, 93-99
    gliding, ground effect, 98-99; photo, 99
    gliding in seabirds, 97-99; illustrations, 97, 98
    gliding, use of thermals in, 95-96; illustration, 95; photos, 96, 97
    hovering, 32, 107-108; photos, 106, 107
    landing, 110-111, 112-113; photos, 112, 113, 114-115
    maneuverability, 108-110; photo, 109
    power-speed curve, 92; illustration, 93
    slow, 103-106; illustration, 104
    soaring, 95-99
    stability, 108
    steering, 108-110
    takeoff, 110-112; photos, 110, 112

Bird wings, 89-92. See also
Feathers
    compared with insects', 47
    flight muscles, 89-90;
        illustrations, 89, 90
    in flapping flight, 99-110;
        illustrations, 102, 103; photos,
        100-101, 104-105
    size, 30
Birds. See also Bird flight; Bird
    wings; Feathers
    bones, 84-85; illustrations, 84
    evolution, 78-81
    flightless, 81-82
    respiratory system, 83-84;
        illustration, 83
    structure and physiology, 82-85
    structure compared with
        insects', 82-83
    warm-bloodedness, value of, 76-
        78
Blériot, Louis, 144
Blowfly, use of energy in flight, 46
Boomerangs as gliders, 118
Borelli, Giovanni, 122, 126
Botfly, deer, 18
Bound vortex, 32-35; illustrations,
    33, 35
    in insect flight, 70-72
Boundary layer, 28-29; illustration,
    29
    and stalling in insect flight, 67
Bourcart's ornithopter, 131;
    illustration, 127
Bow wave, 158
Bullets, 16-17
    firing of, photo, 17
Bumblebees, photos, 30, 60
    temperature control, 44
Butterflies
    admiral, white, photo, 56
    cabbage white, clap-fling-ring
        action, 70-72; photo, 71
    clap-fling action, 70
    coupling of wings, 49
    gliding, 55
    ithomid, photo, 62
    monarch, photo, 46

peacock, photo, 62
silver-washed fritillary, photo,
    75
temperature control, 44
wing-beat frequency and speed,
    60
Canard design, 147; photo, 147
Caudipteryx, 80
Cayley, Sir George, 125-127
Centripetal force and gliding, 95
Chanute, Octave, 138
Chaser, broad-bodied, photo, 63
Chord length of wing, illustration,
    21
Chrysopa, photo, 57
Circulation of airflow, 33, 35; photo,
    35
    in insect flight, 69-74
Clap-fling action
    in bird flight, 107
    in insect flight, 68-70;
        illustration, 70; photo, 69
Clap-fling-ring action in insect
    flight, 70-72; illustration, 71;
    photo, 71
Click mechanism, 51-53
Cockchafer, photos, 49, 66
Concorde, 25, 27
    and transonic drag, 157
    design, 162-165; photo, 164
Condor, wing size, 30
Crane fly, halteres, photo, 50
Curlew, skeleton, illustration, 84
Cyrano de Bergerac, Savinien de,
    theory of dew, 123
Daedalus, 116-118; illustration, 117
Damian, John, 118
Damselflies, photo, 52
    powering of flight, 51
    wing, electron micrograph, 47
    wing-beat frequency and speed,
        60
Danaus plexippus, photo, 46
Danti, Giovanni, 118
Darwin, Charles, 78, 133-134
Da Vinci, Leonardo. See Leonardo
    da Vinci
De Bacqueville, Marquis, 122-123

De Goué, General Resnier, 127

De Groof, Vincent, 130

De Gusmão, Bartolomeu Laurenço, Passarola, 123; illustration, 123

De Havilland, Geoffrey, 155

De Lana de Terzi, Father Francesco, aerial ship, 123; illustration, 122

Degen, Jakob, 127-128

Delta wings of supersonic aircraft, 162-163

D'Esterno, Ferdinand, *Du Vol des Oiseaux*, 133

DH108 Swallow, 155

Dihedral for lateral stability, 28; illustration, 28

Dinosaurs as birds' ancestors, 79-80; illustration, 81

Dive, spiral, 148-149

Dove in flight, photo, 104-105

Downwash, 22

Drag, 19-22, 23-24
form, 20, 21
induced, 21-22, 24, 25
lift-to-drag ratio, 23, 26; illustrations, 22
parasitic, 21
transonic, 157

Dragonflies
ability to hover, 32
control in flight, 63
flight-orientation system, 65
gliding, 55
powering of flight, 51
southern hawker, photo, 37
wing-beat frequency and speed, 60

Du Temple, Félix, 131

EA9 Optimist, photo, 145

Eagle, bald, soaring, photo, 96

Echinomya fera, photo, 62

Elevators, 147, 148; illustrations, 147
trim tab, 149-150; illustration, 149; photo, 149

Elevons, 164

Ellington, Charles, 35, 38-39, 73-74

Elytra, 50; photo, 49

Encarsia formosa, and clap-fling action, 68-69

Energy use in insect flight, 46-47

Engines
gas turbine, 146
high-bypass, 146
jet, 146, 157
piston, 146
ramjet, 167
scramjet, 165
turbofan, 146

Ennos, Roland, 53

F1-11, 162

Feathers, 85-88; illustrations, 87, 90, 91; photos, 77, 86, 87
contour, 87; photo, 86
down, 87
in flapping flight, 102-103; illustration, 103
pennae, 87
plumulae, 87
preening, 87-88
structure, 87; illustrations, 87

Fibrillar muscle, 53-54

Fin for directional stability, 27

Fineness ratio, 20

Fixed-wing inventions, 128-130, 131-143

Flappers. See Ornithopters

Flaps, 151-152, 153; illustrations, 151, 152
Fowler, 151; photo, 151

Flea, photo, 50

Flies
blowflies, use of energy in flight, 46
control in flight, 63-65
dung, yellow, photo, 54
empid, photo, 61
fruit, clap-fling action, 70
horsefly, wing-beat frequency and speed, 60
housefly landing upside down, photo, 67
housefly, takeoff, photo, 65
hover, 32; photos, 31, 72
hover, wing-beat frequency and speed, 60

landing upside down, 66; photo, 67

loss of one pair of wings, 50

parasitic, photo, 62

powering of flight, 51

takeoff, 65

wing-beat cycle, 58-59; illustration, 59

Flight. See also Bird flight; Insect flight

definition, 17

history of theories of, 14-16

in animals, time line, 11

vs. action of balloons and projectiles, 16-17

Flightlessness

in birds, 81-82

in insects, 50; photo, 50

Flow. See Airflow

Flyer, Wright brothers', 143; photo, 142

Forces on body in flight, illustration, 20

Frenulum, 49

Friction, skin, 20-21, 26, 28; illustration, 29

Fritillary, silver-washed, photo, 75

Fuselage for stability, 27

Galilei, Galileo, theory of flight, 16

Galton, Peter, 80

Gecko, flying, photo, 12

Glaucomys sabrinus, photo, 9

Glide ratio, 94

Gliders. See also Hang gliders

early, 120, 126-127, 134-137; illustrations, 121, 135; photo, 126

vs. sailplanes, 146

Gliding. See Bird flight, gliding; Insect flight, gliding

Gossamer Condor, 127

Grasshoppers

powering of flight, 51

temperature control, 43-44

Ground effect in gliding, 98-99; photo, 99

Gull, black-headed, photo, 82

Halteres, 50, 65

Handley Page Ltd., 152

Hang gliders, 133, 135-137, 166-170; photos, 136, 168

control of, 169-170

Hatchetfish, 12

Hawk, sparrow, photo, 91

Helicopter, early concept of, 120

Henson, William Samuel, 128-129

Heron, grey, takeoff, 111

Hesperornis, 81

Hetaerina cruenta, photo, 52

High-bypass engine, 146

Hooke, Robert, 122

Hovering. See Bird flight, hovering; Insect flight, hovering

Hummingbird, hovering, 32, 108; photo, 107

Huxley, Thomas, 79

Hypersonic flight, 165-166; illustrations, 166, 167

Icarus, 116-118; illustration, 117

Ichthyornis, 81

Insect flight, 36-75. See also Insect wings; Insects

articulation of wings, 51-53

bound vortex in, 70-72

clap-fling action, 68-70; illustration, 70; photo, 69

clap-fling-ring action, 70-72; illustration, 71; photo, 71

compared with helicopter flight, 56-58; illustration, 58

control in, 61-65

energetic cost, 46-47

evolution, 36-40

flapping flight, 55-61

fuel for, 46-47

gliding, 55; illustration, 55; photo, 56

hovering, 32, 58

landing, 65-66

leading-edge vortex in, 73-74; illustration, 73

maneuverability, 61-65

nonsteady airflow in, 68-74

orientation in, 65

party-streamer effect, 72-74

powering of, 50-55; illustrations, 51, 53
speed, 59-61; chart, 60
stalling, 66-67
takeoff, 65; photo, 65
temperature control, 43-46
using direct flight muscles, 51; illustration, 51
using indirect flight muscles, 51-52; illustrations, 51, 53
wing movement, illustration, 58
wing-beat cycle, 58-59; illustration, 59
wing-beat frequency, 51, 53-54, 59-61; chart, 60
Insect wings, 47-50
articulation, 51-53
compared with birds', 47
coupling of fore- and hind wings, 49-50; illustration, 49; photo, 48
evolution, theories of, 38-39; illustrations, 38
movement of, 50-55; illustrations, 53
number of pairs, 48-50
scales, photo, 67
structure, 47-50
Insects. See also Insect flight; Insect wings
blood system, 42
early, illustrations, 38, 40; photo (model), 39
factors in success, 40-41
respiration, 43
structure and physiology, 41-43; illustration, 41
structure compared with birds', 82-83
wingless, 50; photo, 50
Instability. See Stability
Iron prominent using clap-fling, photo, 69
Jackdaw, takeoff, photo, 110
Jet engine, 146, 157
Kai Kawus, King, 118
Kelvin, Lord, 139
Kestrel, hovering, photo, 106

Kites
as gliders, 118
delta-wing, 167
Kramer, Melissa, 39
Kukalova-Peck, Jarmila, 38
Lacewing, photo, 57
clap-fling action, 70; photo, 68
Laminar flow, 29; illustration, 29
Langley, Samuel Pierpont, 137-138
Lateral axis of aircraft, 146
Le Bris, Jean-Marie, glider based on albatross flight, 134; illustration, 135
Leading-edge vortex in insect flight, 73-74; illustration, 73
Leonardo da Vinci, 16, 118-121; sketches, 119, 120, 121
LEV (leading-edge vortex), 73-74; illustration, 73
Libellula depressa, photo, 63
Lift, 19-22, 23
Lift-to-drag ratio, 23, 26; illustrations, 22
Lilienthal, Otto, 133, 135-137, 167
Limenitis camilla, photo, 56
Locusts
desert, photo, 64
gliding, 55
Longitudinal axis of aircraft, 146
Macaw, photos, 13, 79
feathers, photo, 77
Mach, Ernst, 156
Mach numbers, 156
MacReady, Paul, 127
Maddox, Dr. Richard, 137
Magnus effect, 33; illustration, 33
Maneuverability
and aspect ratio, 26
and stability, 28
in bird flight, 108-110; photo, 109
in insect flight, 61- 65
Manned flight. See Aircraft
Marden, James, 39
Marey, E.J., "Animal Mechanisms," 107
Markowski, Michael, 167-168
Melolontha melolontha, photo, 49

Microlights, 170; photos, 163, 169, 171

Midge, wing-beat frequency and speed, 60

Millar, Richard, 168

Momentum, 18

Montgolfier, Joseph Michel and Jacques Étienne, 125

Mosquitoes, wing-beat frequency and speed, 60

Moths
    clap-fling action, 70
    coupling of wings, 49-50
    emperor, photo, 45
    hawk, photo, 44
    hawk, speed, 61
    hawk, wing-beat frequency and speed, 60
    noctuid, photo, 15
    pericopid, photo, 20
    temperature control, 44

Mouillard, Louis Pierre, *L'Empire de L'Air*, 134-135

Mozhaiski, Alexander Fedovorich, 131

Musca domestica, takeoff, photo, 65

Nett wing, illustration, 40

Newton, Sir Isaac, 16, 17, *22*

Norberg, Ake, 48

Normal axis of aircraft, 146

Notodonta dromedarius, using clap-fling, photo, 69

Ornithopters, 119-122, 127-128, 130-131; illustrations, 120, 130

Ostrich, wing size, 30

Ostrom, John, 79-80

Owls
    barn, photo, 88
    little, landing, photo, 114-115

P-38 Lightning, 154

P47 Thunderbolt, 154

Panorpa, photo, 42

Parrot, grey, photo, 85

Party-streamer effect
    in delta wings of supersonic aircraft, 163
    in insect flight, 72-74

Parus ater, wing movement, photo, 103

Passarola, 123; illustration, 123

Pectoral muscle of birds, 89-90; illustration, 89

Pelican, white
    in thermal, photo, 97
    using ground effect, photo, 99

Pénaud, Alphonse, 131-132

Pheasants, illustration, 111
    energetic cost of flight, 93
    feathers, photo, 86
    takeoff, 111

Phillips, Horatio, 132-133

Photography, effect on development of aircraft, 137

Pieris brassicae, clap-fling-ring action, 70-72; photo, 71

Pilcher, Percy, 137

Piston engine, 146

Pitching, 146

Planophore, 131

Pocock, George, 118

Prandtl, Ludwig, 28

Preening, 87-88

Pressure, 18-19, 23
    center of, 23; illustrations, *22*
    distribution on airfoil, illustration, 23

Prigent's ornithopter, 131

Projectiles, 16-17, 118

Protopterygotes, 36-38; photo (model), 39

Pterosaur, 12, 80; illustration, 80

Ptychozoon kuhli, photo, 12

Ramjet engine, 167

Ratios
    aspect, 25-26; illustrations, 25
    fineness, 20
    lift-to-drag, 23; illustrations, *22*

Ratites, 81-82

Raven, barrel roll, illustration, 94

Raven microlight, photos, 163, 169

Re (Reynolds numbers), 21

Reading list, 172-173

Reflex in wings for stability, 27; illustration, 27

Resistance, 19

Retinaculum, 50
Reynolds numbers, 21
Robin, energetic cost of flight, 92
Rockets, 16-17
Rogallo, Dr. Francis, 167
Rolling, 146
Rooks, photo, 94
Rudder, 148; illustrations, 148
Saab Gripen, photo, 161
Sailplanes, 146; photo, 145
Sailwings, 168
Saturnia pavonia, photo, 45
Scale (size) and flight capabilities,
    29-30
Scales on moth wing, photo, 67
Scatophaga stercoraria, photo, 54
Sceliphron caementarium, photo,
    59
Scorpionfly, photo, 42
Scramjet engine, 165
Shape of body moving through air,
    19-20. See also Streamlining
Shearing stress, 29
Shock stall, 158-159
Shock waves at transonic and
    supersonic speeds, 156-157, 158;
    illustrations, 156, 158
Sideslipping, 28, 148; illustration,
    28
Sinosauopteryx, 80
Size and flight capabilities, 29-30
Slats, 152-153; photo, 153
Slots, 152; photo, 151
    Handley-Page, 90
Soaring in birds, 95-99
Sound
    barrier, 154-156
    speed of, 18, 156
Speed. See Bird flight, fast;
    Bird flight, slow; Insect flight,
    speed; Sound, speed of;
    Supersonic speeds; Transonic
    speeds; Velocity of airflow
Speed range, 150-151
Spider, money, winglessness, 30
Spin, 153-154; illustration, 154
Spiral dive, 148-149
Spitfire, 154, 158; photo, 159

Spoilers, 152
Squirrel, flying, northern, photo, 9
Stability, 26-28; illustrations, 26, 27,
    28
        and control in early aircraft, 131,
            132, 138, 139
        pilot control of, 139-140
Stalling, 24, 153; illustrations, 24,
    153
    in insect flight, 66-67
    shock, 158-159
Starting vortex, 33, 35; illustrations,
    35
Straight and level flight, speed
    range, 150-151
Streamlining, 20; illustrations, 19
Stringfellow, John, 128-130
Super VC 10, photo, 162
Supersonic aircraft, design
    considerations, 158-165
Supersonic airflow, 155-156
Supersonic speeds, 158
        airfoils suitable for, illustration,
            158
        bow wave, 158
        shock waves, 158; illustrations,
            158
Supracoracoideus muscle of birds,
    89-90; illustration, 89
Surface area of body moving
    through air, 19
Surface-volume relationship, 30;
    illustration, 29
Swallow, photo, 109
    landing, photo, 113
Swan
    landing, photo, 112
    wing size, 30
Sweepback in wings, 27, 160;
    illustrations, 27, 160
Swift, illustration, 111
    energetic cost of flight, 92-93
    takeoff, 111-112; photo, 112
Syrphus balteatus, photo, 31
Tabs, trim, 149-150
Tailplane for longitudinal stability,
    27; illustration, 27
Tarsal reflex, 65

Thermals and gliding in birds, 95-96; illustration, 95; photos, 96, 97
Thrust, 21
Time line of animal flight, 11
Tit, coal, wing movement, photo, 103
Tower jumpers, 118, 121; illustration, 118
Trailing vortex, 24; illustrations, 24, 25
Transonic drag, 157
Transonic speeds, 156-157
    shock waves, 156-157; illustration, 156
Trimming, 149-150
TU 160 Blackjack, 162
Tuck-under, 158
Turbine engine, 146
Turbofan engine, 146
Turbulence, 20, 25, 29
Typhoon, 154
Vanessa io, photo, 62
Variable-geometry wings of supersonic aircraft, 162; illustration, 163
Velocity of airflow, 19
Vertical axis of aircraft, 146
Virgil, *Aeneid*, 106
Viscosity of air, 28
Vortex generator, 160-162; photo, 163
Vortices
    bound, 32-35; illustrations, 33, 35
    bound, in insect flight, 70-72
    leading-edge, in insect flight, 73-74; illustration, 73
    starting, 33, 35; illustrations, 35
    trailing, 24; illustrations, 24, 25
    wingtip, 24, 25; illustration, 24
Vultures, photos, 93, 140
    as inspiration for early aircraft, 134-135, 139
    gliding, 94
    soaring, 95, 96-97
Wagner effect, 35
Wallis, Dr. Barnes, 162

Wasps
    coupling of wings, 50
    mud dauber, photo, 59
    parasitic, and clap-fling action, 68-69
    wing-beat frequency and speed, 60
Weathercocking, 27
Weis-Fogh, Torkel, 35, 68-70, 107
Wenham, Francis, 132
Whittle, Sir Frank, 157
Willughby, Francis, 121-122
Wing taper, 25; illustration, 25
Wing-loading, 26
Wing-warping, 139-140; illustration, 139
Wingless insects, 50; photo, 50
Winglets of early insects, 38-39
Wings, 22-26. See also Bird wings; Insect wings
    chord length, illustration, 21
    high, in monoplanes, for lateral stability, 28
    of supersonic aircraft, 160, 162-163; illustrations, 160, 163
    parts of, illustration, 21
Wingtip vortex, 24, 25; illustration, 24
Wootton, Dr. Robin, 38-39, 47
Worcester, Marquis of, 121
Wren, photo, 92
Wright, Wilbur and Orville, 138, 139-143, 144
    Flyer, 143; photo, 142
Xyloplanes pluto, photo, 44
Yawing, 146; illustration, 149
Yeager, Chuck, 155

All photography by Stephen Dalton, with the following exceptions:

The Aviation Picture Library, London, England, pages 145, 147, 150, 159, 164 (also on back cover)

Dr. David Newman, page 47

Science Museum/Science and Society Library, London, England, pages 126, 136 (also on page 7), 142

Mark Wagner, Aviation Pictures, pages 151, 161

All color diagrams by George Hayhurst.

The illustrations on the following pages appear courtesy of:

Science Museum/Science and Society Library, London, England, pages 117, 119, 120, 121, 122, 123, 124, 129

Dr. Robin Wooton, page 38